Perspectives on Development in the Middle East and North Africa (MENA) Region

Series editor

Almas Heshmati, Sogang University, Seoul, Korea (Republic of)

This book series publishes monographs and edited volumes devoted to studies on the political, economic and social developments of the Middle East and North Africa (MENA). Volumes cover in-depth analyses of individual countries, regions, cases and comparative studies, and they include both a specific and a general focus on the latest advances of the various aspects of development. It provides a platform for researchers globally to carry out rigorous economic, social and political analyses, to promote, share, and discuss current quantitative and analytical work on issues, findings and perspectives in various areas of economics and development of the MENA region. Perspectives on Development in the Middle East and North Africa (MENA) Region allows for a deeper appreciation of the various past, present, and future issues around MENA's development with high quality, peer reviewed contributions. The topics may include, but not limited to: economics and business, natural resources, governance, politics, security and international relations, gender, culture, religion and society, economics and social development, reconstruction, and Jewish, Islamic, Arab, Iranian, Israeli, Kurdish and Turkish studies. Volumes published in the series will be important reading offering an original approach along theoretical lines supported empirically for researchers and students, as well as consultants and policy makers, interested in the development of the MENA region.

More information about this series at http://www.springer.com/series/13870

Khazal Abdullah Auzer

Institutional Design and Capacity to Enhance Effective Governance of Oil and Gas Wealth: The Case of Kurdistan Region

Springer

Khazal Abdullah Auzer
Kurdistan Regional Government
Kiel
Germany

ISSN 2520-1239 ISSN 2520-1247 (electronic)
Perspectives on Development in the Middle East and North Africa (MENA) Region
ISBN 978-981-10-4517-2 ISBN 978-981-10-4518-9 (eBook)
DOI 10.1007/978-981-10-4518-9

Library of Congress Control Number: 2017937120

Printed on acid-free paper

This Springer imprint is published by Springer Nature
The registered company is Springer Nature Singapore Pte Ltd.
The registered company address is: 152 Beach Road, #21-01/04 Gateway East, Singapore 189721, Singapore

Foreword

The potential for natural resource-led development to promote economic growth has been questioned since the 1990s. Between 1970 and 1990, many developing countries in East Asia reduced their economic dependence on primary commodity revenues through growth in manufactured exports. By contrast, the oil-exporting countries in North Africa and the Middle East were unable to benefit from the potential of petroleum wealth to drive their economic development, and today their economies remain heavily dependent on petroleum revenues.

However, the emergence of new oil- and gas-producing states and regions, such as Uganda, Mozambique, Ghana, Papua New Guinea, Suriname, South Sudan and Iraqi-Kurdistan, is triggering renewed debate about the potential for petroleum wealth to drive economic development. This has driven the current research to question what are the specific challenges associated with petroleum-led economic development, and what are the potential constraints? More importantly, however, under what conditions can oil- and gas-rich countries channel more of their revenues into social capacity building, and how can they promote sustainable economic development?

This book presents a 'critical reappraisal' of the so-called resource curse and extends the analysis to consider the political and social dimensions, and thus the importance of the structure of the petroleum sector's governance model. Research that has sought to explain the reasons behind the resource curse suffers from certain limitations. Much of the resource curse literature consists only of cross-country quantitative analysis, which is limited by the objectivity of results, data availability and quantification of variables, such as institutional quality and economic growth, and may not offer sufficiently robust explanations. Other research has used detailed case studies that suffer from limitations of the generalisability of their findings to the study population or community (Matveev 2002; Goldstein and Spiegelhalter 1996; Stiglitz et al. 2009). Therefore, this research adopts a triangulation approach, using a cross-country quantitative data (econometric analysis), a cross-country qualitative research (comparative case study analysis) and a single case study (inductive thematic analysis) to examine issues from different perspectives. Methodologically, this research engages critically with the resource curse literature.

It also contributes to knowledge by investigating the causal factors that may promote or hinder the effective management of oil and gas resources in the Kurdistan Region, which also has implications for the security of the wider region and for global energy security. The project also seeks to generate lessons and policy guidelines to help inform other new petroleum exporting countries and regions about how best to manage their new-found wealth. This book is the most systematic analysis of management of the oil and gas sector in Iraqi-Kurdistan to date and it is suitable for audiences from academia, NGOs, policy makers, and stakeholders (private and national oil and gas companies and Ministries and consultants) in oil- and gas-exporting countries.

Michael Bradshaw
Professor of Global Energy
Warwick Business School
Coventry, UK

References

Goldstein H, Spiegelhalter DJ (1996) League tables and their limitations: statistical issues in comparisons of institutional performance. J Roy Stat Soc Ser A (Statistics in Society) 159(3):385–443

Matveev AV (2002) The advantages of employing quantitative and qualitative methods in intercultural research: practical implications from the study of the perceptions of intercultural communication competence by American and Russian managers. Bull Russ Commun Assoc Theory Commun Appl Commun 1:59–67

Stiglitz J, Sen A, Fitoussi JP (2009) The measurement of economic performance and social progress revisited: reflections and overview. Commission on the Measurement of Economic Performance and Social Progress, Paris, France

Acknowledgements

I would especially like to thank Prof. Michael Bradshaw, Prof. Almas Heshmati, Dr. Ana Galvao for their valuable suggestions for improving my book. Their insightful advice, helpful guidance, inspirational criticism, and patient encouragement assisted the writing of this book in numerous ways. I also would like to take this opportunity to express my gratitude to the Ministry of Natural Resources, the Parliament of the Kurdistan Region, and all those who have been influential in the preparation and completion of this book.

I dedicate this book to the memory of my father, Abudllah Auzer Hostani, who firmly established in me a sense of moral and ethical thinking, to my mother who gave me the love of life, and to my beloved Kurdistan.

Contents

List of Figures

List of Tables

Chapter 1
Overview

Abstract The large and growing body of resource curse literature has used either quantitative or qualitative cross-country approaches to investigate the determinant factors contributing to poor economic performance in natural resource-rich countries, especially petroleum-driven economies, such as Kurdistan Region. Few research studies have used mixed methods to study the resource curse in order to gain a deeper understanding of the challenges facing petroleum-exporting countries that have been unable to convert their petroleum wealth into long-term sustainable development. Author offers different methods to explore the economic, political, and social channels behind the resource curse theory. This chapter explains mixed method study, which consists of three phases: a quantitative cross-country (econometric) analysis, a qualitative cross-country (comparative) policy analysis, and a qualitative case study (semi-structured interviews).

Keywords Resource curse · Petroleum exporters · Petroleum governance · Kurdistan Region · Econometric analysis · Policy analysis · Interview analysis

1.1 Introduction

This work builds on earlier studies that have espoused the notion of the 'resource curse'. The main idea behind the 'resource curse' theory is that natural resource-abundant countries tend to have lower rates of economic growth than those with fewer natural resources (Mahdavy 1970; Gelb 1989; Auty 1993; Sachs and Warner 1995, 1997, 2001; Gylfason 1999).

Various economic theories have sought to explain the 'resource curse' phenomenon, including the 'Dutch disease model', which refers to the potentially negative impacts on the rest of the economy of natural resource windfalls and accompanying rises in exchange rates. The economic transmission channels of the resource curse offered by economists in petroleum exporting countries range from oil price volatility on the international market making oil revenues variable and

K. Abdullah Auzer, *Institutional Design and Capacity to Enhance Effective Governance of Oil and Gas Wealth: The Case of Kurdistan Region*, Perspectives on Development in the Middle East and North Africa (MENA) Region, DOI 10.1007/978-981-10-4518-9_1

uncertain, to the existence of oil revenue-dependent economies and a lack of diversification in the wider economy (Cordon and Neary 1982; Dées et al. 2008; Gelb 2011).

In contrast to economic explanations of the resource curse that have tended to overlook the role of institutional quality and effective management of natural resources, political economy approaches are now paying particular attention to the role of the quality of institutions and the potential interaction between natural resources and institutional quality. The political economy theory of the resource curse explains that the root of poor economic performance in resource-abundant countries lies in the fact that governments have failed to adopt the desired economic policies to ensure effective resource revenue management (Ross 2012; Karl 1999; Watts 2004). In other words, policy makers play a major role in restricting growth through political interventions, rent seeking and corruption.

This book takes a broader social perspective to shed light on the main challenges facing petroleum-producing countries in achieving desirable outcomes. More specifically, the purpose of this research is to develop new insights and to deploy them to produce policy guidelines for the Kurdistan Regional Government (KRG), as a new petroleum-producing region, to enable improvement of the management of petroleum revenue.

1.2 The Kurdistan Region

The autonomous Kurdistan Region in northern Iraq, which emerged in 1991, developed its autonomous institutions, including a parliament and an oil and gas sector, following the fall of the Baath party in the wake of the US-led invasion of Iraq in 2003.

As can be seen in Fig. 1.1, the Kurdistan Region is bordered by Syria to the west, Iran to the east and Turkey to the north. Erbil is its capital city, and Kurdish and Arabic are the official languages of the Region. In addition to Kurds, a diverse collection of ethnic and religious minority groups live side by side in the Kurdistan Region, including Arabs, Turks, Chaldeans, Syriacs, Assyrians, Yizidis, Kakayi and Shabaks. Kurdistan is made up of the four governorates of Erbil, Slemani, Deuhok and Halabja. In addition to these governorates, the KRG governs parts of Nineveh and Diyalah governorates, and also claims jurisdiction over a wider set of 'disputed territories', including Kirkuk province.

1.2.1 The Kurdistan Region's Political System

In 2005, the Kurdistan Region was officially recognised in the Iraqi constitution as a federally semi-autonomous political region. Its three main institutions are the Kurdistan Regional Government (KRG), the Kurdistan Parliament, which was

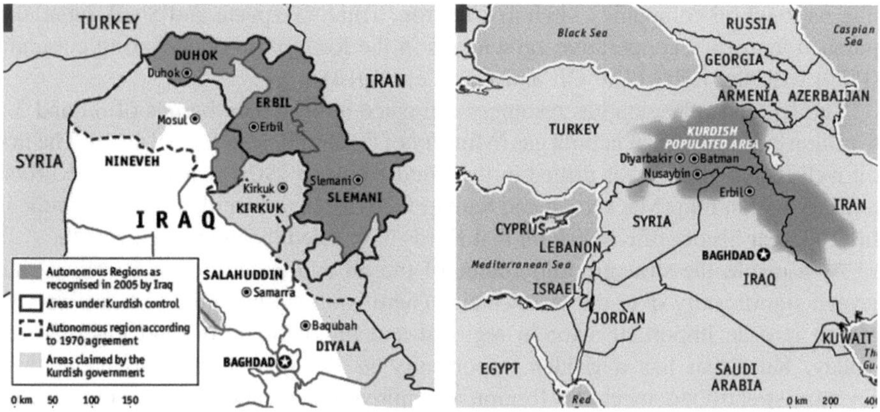

Fig. 1.1 Kurdistan Region, Iraq. *Source* Google (2015)

established in 1992, and the Kurdistan Region Presidency, established in 2005. The Kurdistan Region's parliament has 111 seats and ratified the proposed constitution in June 2009. The last parliamentary elections were held on 21 September 2013. According to Iraq's federal constitution, the Kurdistan parliament has considerable power to debate and legislate on policies in a wide range of areas, including health services, education and training, policing and security, the environment, natural resources, agriculture, housing, trade, industry and investment, social services and social affairs, transport and roads, culture and tourism, sport and leisure, and ancient monuments and historic buildings (Kurdistan Regional Government 2015). When the Islamic State of Iraq and Syria (ISIS) seized control of vast areas of Iraq, several governorates and their respective authorities suffered total collapse. The central government's loss of political, financial and military power and dysfunctional federal institutions have dramatically strengthened the KRG's authority over foreign affairs and operational areas, since the military forces of the Kurdistan Region (Peshmerga) have been defending a largely flat and naturally defenceless frontier against terrorist attacks by forces loyal to ISIS. Furthermore, the Peshmarga has taken control of most disputed territories, such as the giant oil-rich fields in the city of Kirkuk (Solomon 2014; The Guardian 2014).

1.2.2 The Kurdistan Region's Oil and Gas Sector

The peace and relatively high security of the Kurdistan Region compared with the rest of Iraq has enabled the KRG to develop its oil sector and lease much of its land, on the basis of production sharing agreements, to energy production companies for international resource exploration (see Fig. 1.2). As an example of the sector's attractiveness, Exxon Mobile signed a contract in 2011 for six exploration blocks, and in doing so relinquished access to blocks in southern Iraq. Other well-known international oil companies, such as Chevron, Total, Gazprom and Shell, have also pursued their own cooperative agreements in the form of product sharing contracts (PSC) with the KRG (The Oil and Gas Year 2014).

The Kurdistan Region has resources estimated at 45 billion barrels of oil and 3–4 trillion cubic metres of natural gas (Ministry of Natural Resources 2014a, b). The first oil well (Chya Surkh) was drilled in the Middle East in 1901, which was in Kurdish territory within Iraq (Mackertich and Samarrai 2015). However, this is the first time in the 100 year history of Kurdish oil that Kurds have ruled their endowed wealth.

As a result, the strategic importance of the oil and gas sector in Kurdistan has grown significantly over the last decade. There is now a potential to transform the region into an important actor in regional and international oil and gas markets. Today, Kurdistan has a golden opportunity to use this strength to create new revenue streams, to meet the Region's employment creation needs, and to contribute to the social welfare of its people. Therefore, this book studies how the KRG can best manage the economic, political and social challenges that it faces, so that the abundance of natural resources can be a 'blessing' rather than a 'curse' for Iraqi-Kurdistan.

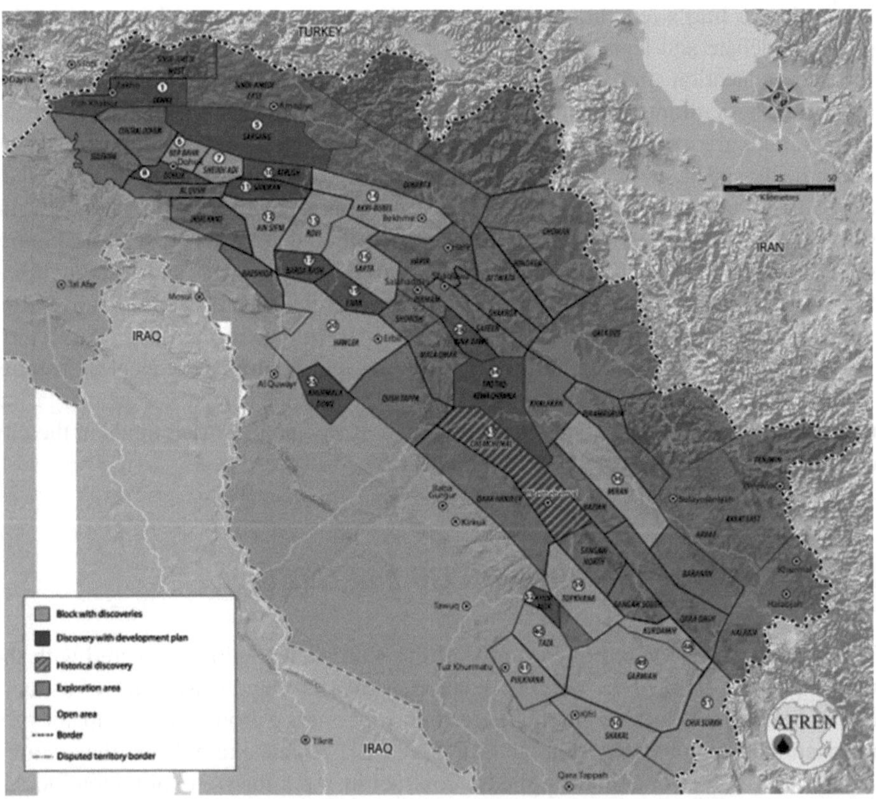

Fig. 1.2 Discoveries and development of 'blocks' in the Kurdistan Region. *Source* Ministry of Natural Resources (2015)

1.3 Aims and Objectives

The aims of this book are:

1. To understand why some petroleum-rich countries succeed in developing their economies whilst others fail.
2. To devise a regime for the effective governance of Iraqi-Kurdistan's oil and gas wealth.

These aims are achieved by pursuing the following objectives:

1. To examine the causes and consequences of the 'resource curse' and identifying the major policy challenges associated with oil and gas wealth.
2. To analyse the economic performance, practices and experiences of a number of oil and gas rich countries.
3. To explain the development pattern of the oil and gas sector in the KRG and in the wider economy.
4. To evaluate current governance structures relating to the oil and gas sector in the KRG.
5. To devise a management structure and policy framework for the efficient and transparent management of Iraqi-Kurdistan's oil and gas wealth.

1.4 Methods of Study

This section provides a summary of the approach taken to the current research. The present research is designed to offer an explanation for the success or failure of resource-abundant countries to derive benefits from the large revenues accruing from natural resources, using explanatory theories of the resource curse drawn from the extant literature.

Much of the resource curse literature consists only of cross-country quantitative analysis, which is limited by the objectivity of results, data availability and quantification of variables, such as institutional quality and economic growth, and may not offer sufficiently robust explanations. Other research has used detailed case studies that suffer from limitations of the generalisability of their findings to the study population or community (Matveev 2002; Goldstein and Spiegelhalter 1996; Stiglitz et al. 2009). Therefore, this research adopts a triangulation approach, using a cross-country quantitative data (econometric analysis), a cross-country qualitative research (comparative case study analysis) and a single case study (inductive thematic analysis) to examine issues from different perspectives.

A key strength of the mixed methods approach is that it avoids the weakness of applying a single quantitative or qualitative approach alone, allowing the researcher to benefit from the strengths of each approach in order to provide a comprehensive understanding of the phenomenon (Tashakkori and Teddlie 1998; Creswell and Plano Clarke 2007). The mixed methods approach used to collect and interpret data on the resource curse phenomenon is designed to elicit an accurate representation of reality (Foss and Ellefsen 2002). The combination of multiple methods in a single study adds depth and breadth to the investigation. That is, mixed methods provide in-depth understanding of complex phenomena such as the resource curse. In addition, the current research has a unique rationale based on the research questions and the aims of the research.

For the current research, a sequential mixed methods approach was adopted. A three-phase, sequential mixed methods approach incorporating both quantitative and qualitative approaches was followed. Phase one was a quantitative cross-country analysis, using econometric methods to achieve Objective 2: to reappraise the importance of the key economic, political, and social challenges facing resource-rich countries. This was followed by a qualitative cross-country case study, using comparative analysis to analyse the economic performance, practices and experiences of a number of oil- and gas-rich countries to achieve Objective 3. Finally, a qualitative single case study approach was adopted to explore the economic, political, and social challenges facing the Kurdistan Region's oil and gas sector, thereby achieving Objectives 4 and 5. This combination helps to emphasise the strengths and minimise the weaknesses of both quantitative and qualitative approaches across the whole research (Tashakkori and Teddlie 1998; Creswell et al. 2007). The findings of this research contribute to the body of theoretical, methodological and practical knowledge.

The current research adopted a pragmatic paradigm, as most methodological researchers (e.g. Teddlie and Tashakkori 2003; Creswell et al. 2007) have proposed a pragmatic paradigm for a mixed methods approach. The pragmatic paradigm supports understanding of problems in the "real world". In the case of the current research, this paradigm was oriented towards understanding the reasons behind the effective governance of oil and gas wealth, and the researcher, as a pragmatist,

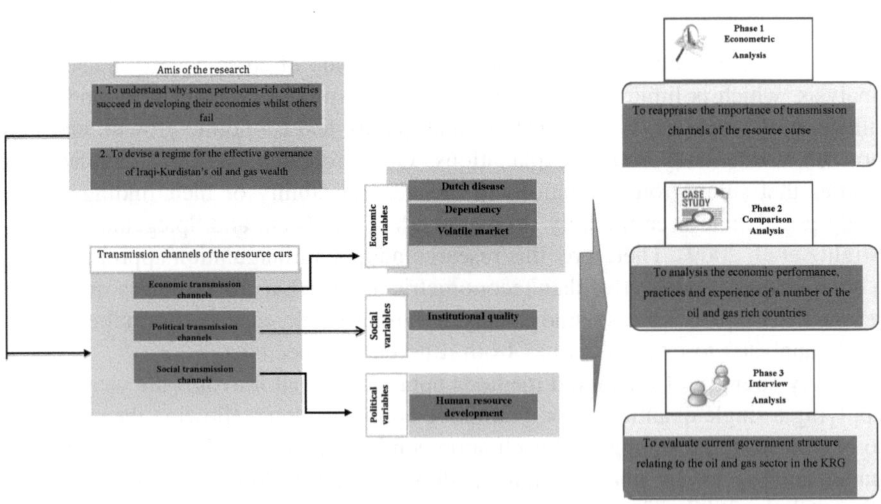

Fig. 1.3 Overview of research design

decided to use a range of tools to investigate comprehensively and from multiple perspectives the challenges facing petroleum governance. As a result, the interpretation of data involves both deductive and inductive reasoning using a mixed methods design. Figure 1.3 provides an overview of the research and its three sequential phases.

1.5 Book Outline

This book consists of seven chapters. The first chapter provides an introduction to the study. It presents the rational of this research, its aims and objective and the study methods. It ends with an outline of the book.

In order to identify the major challenges facing resource-rich countries, a comprehensive literature review was conducted and is presented in Chap. 2. On the basis of this, the existing challenges were divided into three categories: economic, political and socio-economic. In addition, these different transmission channels of the resource curse were re-tested through econometric analysis. Chapter 3 re-appraise the importance of economic, political and social transmission mechanisms of the resource curse employing both cross-sectional and panel data methods. Particular attention has been given to the role of institutions and the capacity of human resources within petroleum-based economies. Chapter 4 presents a cross-country case analysis (Norway, Kuwait, Azerbaijan and Nigeria) to identify effective economic, political and socio-economic policies that address the various resource curse challenges. Chapter 5 presents an overview of Kurdistan's petroleum development through publicly available documents. Chapter 6, which is based on the findings of a series of semi-structure interviews, identifies the challenges surrounding the Kurdistan Region's oil and gas sector. Chapter 7 presents a discussion of the specific implications of the research findings for both the KRG and other new oil-and gas-producing states. It concludes by placing the findings of the study in the wider academic context of research on the 'resource curse'.

References

Auty RM (1993) Sustaining development in mineral economies: the resource curse book. Routledge, London

Cordon WM, Neary JP (1982) Booming sector and deindustrialisation in a small open economy. Econ J 92(368):825–848

Creswell JW, Hanson WE, Plano Clark VL, Morales A (2007) Qualitative research designs: Selection and implementation. The Counselling Psychol 35(2):236–264

Dées S, Gasteuil A, Kaufmann RK, Mann M (2008) Assessing the factors behind oil price changes. Working Paper 855, European Central Bank, Frankfurt, Germany

Foss C, Ellefsen B (2002) The value of combining qualitative and quantitative approaches in nursing research by means of method triangulation. J Adv Nurs 40(2):242–248

Gelb A (1989) Oil windfalls: blessing or curse? Oxford University Press, Oxford

Gelb A (2011) Economic diversification in resource rich countries. Mimeograph, Center for Global Development, Washington, DC
Gylfason T (1999) Natural resources and economic growth: a Nordic perspective on Dutch disease. UNU/WIDER Working Paper No. 167, Helsinki, Finland
Goldstein H, Spiegelhalter DJ (1996) League tables and their limitations: statistical issues in comparisons of institutional performance. J Roy Stat Soc Ser A (Statistics in Society) 159 (3):385–443
Karl TL (1999) The perils of the petro-state: reflections on the paradox of plenty. J Int Aff 53 (1):31–38
Kurdistan Regional Government (2015) Available at: http://cabinet.gov.krd
Mackertich DS, Samarrai AI (2015) History of hydrocarbon exploration in the Kurdistan Region of Iraq. GEOARABIA 20(2):181–220
Mahdavy H (1970) The patterns and problems of economic development in rentier states: the case of Iran. In: Cook MA (ed) Studies in the economic history of the middle-east. Oxford University Press, Oxford, pp 428–467
Matveev AV (2002) The advantages of employing quantitative and qualitative methods in intercultural research: practical implications from the study of the perceptions of intercultural communication competence by American and Russian managers. Bull Russ Commun Assoc Theory Commun Appl Commun 1:59–67
Ministry of Natural Resources (2014a) Production report. Kurdistan Regional Government, Ministry of Natural Resources, Erbil, Iraq. Available at: http://mnr.krg.org
Ministry of Natural Resources (2014b) Revenue report. Kurdistan Regional Government, Ministry of Natural Resources, Erbil, Iraq. Available at: http://mnr.krg.org
Ministry of Natural Resources (2015) Oil production, export, and consumption report. Kurdistan Regional Government, Ministry of Natural Resources, Erbil, Iraq
Ross ML (2012) The oil curse: How petroleum wealth shapes the development of nations. Princeton University Press, London
Sachs DJ, Warner AM (1995) Natural resource abundance and resource growth. Working Paper 5398, National Bureau of Economic Research, Cambridge, MA. Available at: http://www.nber.org/papers/w5398. Accessed 12 Feb 2016
Sachs DJ, Warner AM (1997) Natural resource abundance and economic growth. Research Paper, Centre for International Development and Harvard Institute for International Development, Cambridge, MA
Sachs JD, Warner AM (2001) The curse of natural resources. Eur Econ Rev 45(4):827–838
Solomon E (2014) Kurdish forces seize oilfields near Kirkuk. The Financial Times, 12 July. Available at: http://www.ft.com/cms/s/0/d7ebbc7c-09af-11e4-86f9-00144feabdc0.html #axzz3zw63544a. Accessed 12 Feb 2016
Stiglitz J, Sen A, Fitoussi JP (2009) The measurement of economic performance and social progress revisited: reflections and overview. Commission on the Measurement of Economic Performance and Social Progress, Paris, France
Tashakkori A, Teddlie C (1998) Mixed methodology: combining qualitative and quantitative approaches. Sage Publications, Thousand Oaks, California
Teddlie C, Tashakkori A (2003) Major issues and controversies in the use of mixed methods in the social and behavioral sciences. In: Tashakkori A, Teddlie C (eds) Handbook of mixed methods in social & behavioral research. Sage Publications, Thousand Oaks, California, pp 3–50
The Guardian (2014) Iraq government and Kurdish region strike budget and oil export deal. *The Guardian*, 2 December. Available at: http://www.theguardian.com/world/2014/dec/02/iraq-government-kurdish-budget-oil-export-deal. Accessed 12 Feb 2016
The Oil and Gas Year (2014) The Kurdistan region of Iraq 2014. United Arab Emirates, Dubai
Watts M (2004) Resource curse governmentality, oil and power in the Niger Delta, Nigeria. Geopolitics 9(1):50–80

Chapter 2
Challenges in Petroleum Rich Countries

Abstract This chapter identifies the main channels through, which oil and gas resources may promote or impede economic growth from a broader social perspective. In addition to the economic and political channels of the resource curse, socio-economic challenges, such as a low level of human capacity building, may exacerbate the adverse effect of petroleum resources on long-term economic growth. Petroleum states under-invest in education and workforce skills because their economies are based on their endowments of petroleum resources. A key shortcoming of the resource curse literature is the lack of clear understanding of the effect of the managerial model of the oil sector, in particular its effects on the sector's economic performance. Few case studies have focused on effective administrative design as a causal factor affecting the performance of the oil sector.

Keywords Resource curse · Economic challenges · Political economy challenges · Social challenges · Managerial model of petroleum sector

2.1 Introduction

The concept of the resource curse suggests that countries endowed with natural resources, such as minerals, oil and gas, have been less able to develop their economies than others with fewer natural resources (Auty 1993). However, the success stories of some natural resource-abundant countries, such as Norway and Botswana, suggest that abundance of natural resources is not a curse per se, but rather that government mismanagement of natural wealth is to be blamed for the emergence of the resource curse in many resource-rich countries such as Nigeria.

This investigation considers the findings of research in the field, which have identified a diverse set of challenges associated with the natural resource curse faced by hydrocarbon-producing states. Economists have sought to investigate the relationship between vast natural resource revenues and economic performance. Dutch disease, volatility of oil prices, resource revenue-based economies and a lack of diversification in the wider economy are significant economic challenges relating

K. Abdullah Auzer, *Institutional Design and Capacity to Enhance Effective Governance of Oil and Gas Wealth: The Case of Kurdistan Region*, Perspectives on Development in the Middle East and North Africa (MENA) Region, DOI 10.1007/978-981-10-4518-9_2

to the resource curse (Cordon and Neary 1982; Sachs and Warner 2001; Dées et al. 2008; Gelb 2011).

Another focus of the literature is on political explanations for the resource curse. It is considered that weak governance and poor administrative quality may promote rent-seeking activities rather than productivity in resource-rich countries: many 'rentier states' are overwhelmed by corruption and poor human capacity building, resulting in economic under-development. From a political point of view, it is concluded that the dilemma associated with abundant natural resources stems from the fact that decision makers in oil-rich countries have failed to take effective economic measures to combat macro-economic instability caused by the frequent 'boom and bust' of oil cycles. In other words, government policies are key in generating benefits from petroleum resources (Mahdavy 1970; Karl 1999; Watts 2004; Di John 2011; Ross 2012).

Against this background, this chapter is organised into three main sections addressing economic aspects of the 'resource curse', political aspects of the 'resource curse', and the inadequacies of the literature.

2.2 Economic Aspects of the 'Resource Curse'

2.2.1 Dutch Disease

The extraction of natural gas in the Netherlands in the 1960s resulted in a boom in the natural resources sector, and with it emerged a new phenomenon known as 'Dutch disease'. Economists see this as an economic dimension of the wider 'resource curse' and link it to the influx of resource income into newly resource-rich countries. A surge of petro-dollars into a newly resource-rich country leads to appreciation of the national currency. This increases costs for the domestic manufacturing sector and finances the expansion of imports. Consequently, de-industrialisation is a side-effect of Dutch disease (Cordon and Neary 1982).

Since the introduction of the concept, a substantial literature has developed. For example, Cordon and Neary (1982) developed an economic model to explain Dutch disease in terms of three economic sectors: the natural resources sector; the non-resource traded goods sector, consisting of export industries with prices set by competition on world markets; and the non-traded goods sector, consisting of activities for which the domestic market determines prices. They argue that Dutch disease appears in a resource-rich country when the non-traded goods sector is squeezed or 'crowded out' by growth in the other two sectors. They also argue (1982, p. 983) that this may occur through two effects: the 'spending effect' and the 'resource movement effect.' They highlight that, on the one hand, a boom in income from the natural resources sector increases demand for non-traded goods, resulting in higher prices for these goods. This may provoke an increase in the real exchange rate, which may result in a deterioration in the competitiveness of the non-resource traded sector. This is the 'spending effect'. The 'resource movement effect', on the other hand, comes into play when the workforce shifts away from the traded goods sector into the non-traded goods sector owing to increases in wages. Both effects may lead to de-industrialisation in the non-resource traded goods sector. The relocation of labour away from the traded goods sector may also impact on the non-traded goods sector (Cordon and Neary 1982). In this respect, Kuralbayeva and Stefanski (2013) state that the non-traded sector may often be less productive than the manufacturing sector. This may result from the movement of less skilled labour into the non-traded sector and the fact that specialist labour tends to remain in the manufacturing sector.

In addition to the spending and resource movement effects, Dutch disease is also linked to an inappropriate allocation of resources between the traded and non-traded goods sectors that may also result in lower economic growth. According to Sachs and Stiglitz (2007), in cases where the proceeds of an oil boom are invested in the non-traded goods sector, there may be evidence of Dutch disease. In other words, natural resource revenues tend to be consumed by public, non-tradable projects, such as roads, construction, power, telecommunications and other services. The manufacturing sector is squeezed, the exchange rate appreciates as a result of increased prices owing to increased aggregate demand, the price of imports falls, and hence the manufacturing sector becomes uncompetitive in the face of cheaper imports. In a more recent study, Saad-Filho and Weeks (2013) state that the effects of Dutch disease may relate to the failure of economic policy in resource-rich countries, which results in a decline in productivity in the traded goods sector.

In summary, a review of the literature shows that economic dependence on natural resource revenues may pose a major threat to the development of the manufacturing and agricultural sectors in a resource-rich economy. This threat is the result of a lack of adequate investment and financial support, weakening the competitiveness of these industries. Furthermore, Davis and Tilton (2005) state that resource-rich countries should take into account the negative growth effect of natural resource depletion when determining resource income investment policies, particularly in relation to the agricultural and manufacturing sectors. In other words, sustaining the development of the non-resource sector should be an important part of maintaining a diverse economic structure.

Despite evidence that Dutch disease can be attributed to natural resources wealth, there is also evidence of significant heterogeneity in outcomes. For instance, Auty (2001) stresses that resource-rich countries have experienced different economic outcomes following the discovery of resources. While a number of countries, such as Nigeria, Ghana and Mexico, have suffered from challenges associated with their resource wealth, other countries, such as Malaysia, Botswana and Chile, have utilised their resource income to develop more diverse economic productivity. Dutch disease may therefore cause severe symptoms in some countries and less prominent ones or none in others. The key challenge for newly resource-rich countries and regions is to devise an economic strategy that mitigates the risk of Dutch disease.

The issue of heterogeneity of effects has been investigated in several other studies. Ismail (2010) conducted a survey revealing Dutch disease in a number of oil-extracting countries between 1997 and 2004. The findings of this study, consistent with the theory's predictions, confirm that an influx of oil rents results in a decline in the productivity of the manufacturing sector due to the inflow of revenue and the movement of labour and capital in the non-traded sector. The above finding was also found in a seminal study of an earlier period by Sachs and Warner (2001), demonstrating that resource-rich countries failed to grow their manufacturing sectors between 1970 and 1990.

Contrary to the above studies, in a number of other countries, such as Russia, the abundance of resources has resulted in the development of their manufacturing sectors, contrary to the expectations of the Dutch disease hypothesis. Works by Tabata (2012) and Dobrynskaya and Turkisch (2010) are among two recent investigations that show that Russia managed to increase the productivity of its manufacturing sector from 1999 to 2007, despite observed symptoms of Dutch disease. Furthermore, they note that these symptoms may be linked to effects more diverse than those covered by the Dutch disease prediction; hence, Russia may be considered to be a specific case. These authors associate economic growth in Russia with the adoption of a flexible monetary policy and a rent-sharing policy. One example of the monetary policy is the stabilisation of the exchange rate of the Russian currency that resulted in maintaining the competitiveness of a number of domestic products in both domestic and world markets. In this regard, Ickes and Gaddy (2005) argue that 'excess cost' may be regarded as another supportive measure in the form of rent sharing, which may lead to higher productivity in the manufacturing sector. For instance, railway carriage production has increased since oil has been shipped by rail rather than pipeline; however, the transportation of oil by rail is more expensive than by pipeline. This may be considered as an excess cost policy that encourages production in the manufacturing sector (Ickes and Gaddy 2005). As Lipscomb et al. (2010) demonstrate, Indonesia is another resource-abundant country that considerably increased its manufacturing goods production between 2003 and 2008.

In summary, the Dutch disease hypothesis suggests that resource-abundant countries tend to exhibit low economic growth rates because the influx of revenues leads to exchange rate appreciation. This may affect the manufacturing sector through spending effects and resource movement effects, resulting in a decline in the share of manufacturing due to a decrease in competitiveness in both domestic and world markets. However, empirical results suggest heterogeneity in terms of impact on the economic performance of resource-rich countries. In this regard, several other studies have shown that a number of resource-abundant countries have managed to develop their manufacturing sectors, leading to export competitiveness on the world market. In short, factors other than exchange rates may also play a major role in the efficient use of resource revenues (Tabata 2012; Dobrynskaya and Turkisch 2010). For instance, weak institutional capacity may be a negative determinant in managing Dutch disease (Bunte 2011). Therefore, it is necessary to learn from the experiences of countries that have been successful in managing the challenges of resource renting.

2.2.2 *Oil Price Volatility*

Since the 1970s, crude oil prices have been subject to high volatility. Political tension associated with oil-producing countries appears to be a major reason for changes in oil prices. Uncertainty in oil-exporting countries may translate into a decline in oil supply on the international market, which may result in a rise in oil prices. The history of the oil market shows that several events have had a considerable influence on oil prices, including the Iranian revolution in 1980, the invasion of Iraq in 2003, and other political conflicts in African countries such as Nigeria and Venezuela (BP Statistical Review of World energy 2015). Moreover, oil prices may be influenced by OPEC's decisions with respect to oil supply (King et al. 2012). Researchers have also suggested fluctuating demand for crude oil as another factor affecting the oil market that may have an impact on oil prices. Increasing demand from emerging markets, such as China and India, have been a determinant factor in oil price rises since 2003 (King et al. 2012). Sornette (2009) believes that speculation plays a major role in oil price rises and explains that uncertain political situations in oil-producing countries may enhance the oil market, whereas King et al. (2012) state that there is insufficient evidence that speculative behaviour may influence oil prices. Considering the impact of the dollar exchange rate on oil price volatility, Zhang et al. (2008) argue that the changing value of the dollar also plays a significant role in the volatility of oil prices because oil markets trade in US dollars. For example, depreciation in oil prices may lead to oil price increases owing to increasing demand for crude oil. The literature also shows that the refinery sector, as a major oil consumer, may have an effect on oil prices (Dees et al. 2008). These exogenous factors shed light on the fact that oil price volatility may translate into volatile oil revenues, which makes oil revenue management a challenging task for oil-based economies.

Mismanagement of fluctuating oil revenue

Crude oil prices have recently been highly volatile, as shown in Fig. 2.1. Oil prices remained high and relatively stable between 2012 and 2013, while increasing financial activities in oil markets may have been the cause of recent oil price fluctuations (Lombardi and Van Robays 2011; Labban 2010). Although oil prices fell unexpectedly by about 50% in the second half of 2014, from US$108 (bpd) in July 2014 to US$48 (bpd) in October 2015, the key points pressuring the oil market include sluggish demand, ample supply and a strong US dollar. The total oil supply has continued to grow as a result of a spike in US oil supply on the back of shale exploration and the expansion of OPEC exports (Baffes et al. 2015).

Oil companies tend to be most interested in oil trade investment, which may produce greater profits than oil production investment when oil prices rise. In this

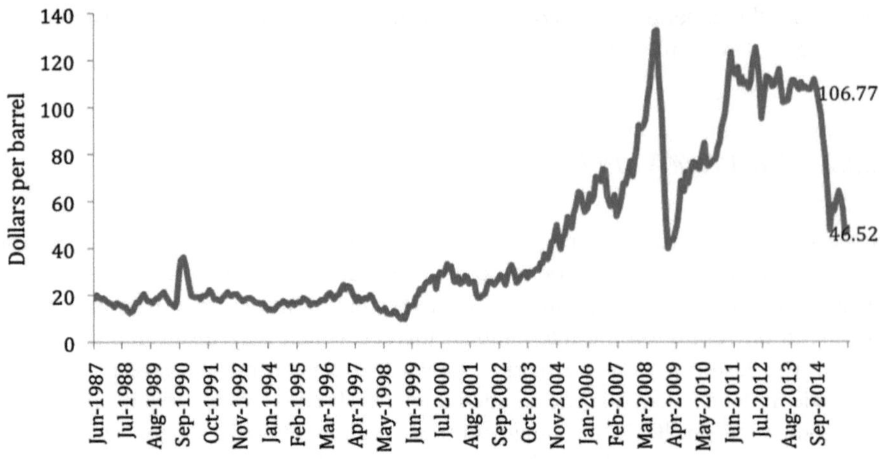

Fig. 2.1 Europe Brent Spot Price FOB May 1987–October 2015. *Source* US Energy Information Administration (2015)

context, oil price volatility may be transmitted to the level of public spending because of the instability of oil revenues available to oil-exporting countries. This volatile fiscal expenditure may have a negative impact on public projects, and may lead to poor economic performance by oil-exporting countries (Lorde et al. 2009; Ramey and Ramey 1995). Hence, the lack of an appropriate fiscal policy associated with the optimal use of oil revenues may have undesirable effects on economic growth, such as macro-economic volatility, low quality of public expenditures and budget deficits (Medas and Zakharova 2009; Sturm et al. 2009; Bacon and Kojima 2008).

Macro-economic volatility may pose a major threat to economic growth. Macro-economic stability may be affected by changes in public spending as a result of 'boom and bust' oil prices. According to Al-Ezzee (2011) and Barrell and Kirby (2008), the expansion of public spending, particularly in infrastructure rather than investment, may make the tradable sector vulnerable to the floating real exchange rate. Moreover, both the lowered competiveness of the tradable sector (Van der Ploeg and Poelhekke 2009) and a lack of adequate investment play major roles in the poor economic performance of oil-exporting states (Al-Marhubi 1998). In addition, a decline in private investment due to oil price fluctuations may have negative effects on economic growth (Cologni and Manera 2013).

Low-quality public expenditure is one explanation for the negative impact of volatile oil revenues on the productivity of the tradable sector. Sturm et al. (2009) argue that rising oil revenues induce excessive social spending and infrastructural investment. In this regard, lack of appropriate investment guidelines often results in inaccurate evaluation and selection of public projects—so-called 'white elephants' or 'cathedrals in the desert'. This may cause low efficiency and productivity of fiscal policy in oil-producing countries. Medas and Zakharova (2009) also highlight that some oil-producing countries, such as Nigeria and Algeria, have allocated oil revenues to ambitious public projects without economic return, in pursuit of political objectives. More generally, raising public expenditure may lead to imbalances in the state budget.

Oil revenue volatility also makes balancing the budget a challenge for oil-producing countries. El Anshay and Bradley (2012) explain that budgets may be in deficit after a drop in oil revenues. Such budget deficits occur because of excessive public spending in periods of rising oil revenue, which may result in the state having to borrow in order to offset the deficit. This may hamper economic progress by leading to the discontinuation of projects associated with infrastructure development or economic diversification that aim to expand sources of revenue to counter dependence on oil revenues. Similarly, Van Wijnbergen and Budina (2011) stress that oil revenue volatility is likely to cause oil-exporting states to fall into debt. This may occur because of a lack of a sustainable fiscal policy to avoid budgetary deficit once oil prices fall. This would come into play as a result of the mismanagement of oil revenues in oil-producing countries.

Managing oil revenue volatility

In dealing with all the above-mentioned challenges, oil-exporting countries may implement a number of fiscal measures in order to utilise their oil income effectively. Managing oil revenues by setting an annual budgetary resource on the basis of a conservative estimate of oil prices is a common fiscal policy among oil-producing countries (Ossowski et al. 2008; Sturm et al. 2009). In addition, oil-exporting states may attempt to smooth public expenditure and mitigate the risk of budget deficit by adopting conservative oil price assumptions in budgets. Sturm et al. (2009) argue that the use of conservative budget-based oil prices may pose a major threat to fiscal transparency because authorities may have more room to manoeuvre in spending surplus revenues. For example, in Saudi Arabia, public expenditure has increased by 15–20% in recent years, despite its budget being based on conservative estimations of oil prices. This is likely to be a result of increased distribution-related policies, such as subsidies and public sector jobs, to counteract increasing social pressures. Similarly, Ossowski et al. (2008) note that such fiscal policy may have drawbacks associated with the effective and transparent use of oil revenues. Dudley (2011) explains that high budget price estimates may make fiscal policy more vulnerable when real oil prices are lower than oil price estimates. For instance, the oil exporting countries since mid-2014 have experienced a sharp decline in oil revenues. In the current low level oil price environment it is clear that only countries with well-designed institutional fiscal polices, such as Norway, are able to cope with the significant fall in oil rents (Arezki and Blanchard 2015).

Another institutional mechanism is a 'stabilisation fund', which saves a proportion of oil revenues when oil prices increase in order to combat the consequences of oil revenue volatility (and Dutch disease). Reserving a proportion of oil revenues may serve to restrain excessive public spending, and thus real exchange rate rises and inflation, during periods of rising oil prices (Ossowski et al. 2008;

Davis et al. 2003; Sturm et al. 2009). Furthermore, a stabilisation fund may serve to adjust budget deficits on 'rainy days' when oil prices fall. Researchers also note that some oil-exporting countries may build up oil funds, such as savings funds, in order to set aside a part of their oil income for future generations because of the exhaustible nature of oil revenues. They argue that a number of factors may make the successful management of oil funds challenging for many oil-producing countries. These factors include low institutional capacity, and a lack of transparency, accountability and political commitment.

Explicit fiscal rules may enhance the quality of economic institutions. Transparent and accountable institutions lead to the restriction of political interventions in terms of managing natural resource revenues. Norway and Botswana have been able to manage their natural resource rents by introducing effective institutional policies (Sturm et al. 2009; Ossowski et al. 2008).

Heterogeneous experiences of institutional fiscal policy

Research shows heterogeneous experiences in resource-rich countries where attempts have been made to establish resource funds to mitigate the negative effects of oil revenue volatility on economic performance. Therefore, reasons for the success or failure of a particular resource-rich country are debatable. Norway's success in resource management, for instance, derives from a variety of factors. The imposition of explicit fiscal rules for oil funds may be the major reason for Norway's success in its management of natural resources (Sturm et al. 2009). Moreover, monetary policy, with inflationary and exchange rate targets, transparency, high levels of human capacity and infrastructure, and good governance are also important in explaining the high performance of Norwegian resource funds.

According to Benedictow et al. (2013), Russia offers another example of the successful implementation of institutional fiscal policies, addressing the negative effects of oil price volatility on macro-economic instability. In addition to saving a part of its oil revenues in a sovereign wealth fund, and preventing high inflation and real currency appreciation through the central bank, the Russian government has played a major role in smoothing public expenditure and productivity in the non-oil sector. However, Sturm et al. (2009) highlight that public expenditure increased in Russia in 2007–2008, and the expansion of public spending was a contributory factor in the results of the parliamentary and presidential elections in 2007. This resulted in the use the savings fund to adjust the budgetary deficit. The fund was further depleted in the aftermath of the 2008 global financial crisis, but has since been replenished. However, the constant decline in oil prices since the second half of 2014 has resulted in drawdown in the reserve fund to adjust the budget deficit (Fouche and Milhench 2015).

On a much smaller scale, Ossowski et al. (2008) note that Timor-Leste has been running its oil fund positively, as it has been well integrated into the budget system and managed within an overall fiscal framework. This study also finds that transparency and accountability are further reasons for the success of Timor-Leste's oil fund management. Similarly, Davis et al. (2003) emphasise that oil funds may operate effectively if they are fully integrated into the central government's budgetary process. This leads to enhancement of transparency and accountability relating to oil revenue management because all transfers into or out of the fund go through the central government budget and are reported by the government. For example, a study by Usui (2007) shows that Azerbaijan has failed to manage its oil fund effectively. This may be attributed to the design of and fiscal rules associated with its oil fund. Azerbaijan's fund is separate from the public budget, so savings can be invested in public projects outside the budget plan. This may result in public expenditure increases and greater opportunities for political intervention.

Conversely, Indonesia and Malaysia have been able to manage their oil revenues effectively through appropriate investment policies associated with diversification and industrialisation of the economic structure (Coutinho 2011). In addition, efficient fiscal rules, which promote budget balance, investment in the non-tradable sector and human resource development, and high institutional capacity play major roles in managing resource revenues efficiently in both countries (Coutinho 2011).

In summary, the empirical studies reviewed here show that oil revenue-based fiscal policies play a significant role in addressing the imbalances in public revenues faced by oil-exporting countries due to the highly volatile nature of oil prices. The major purpose of sound and effective fiscal policies is to enhance the economic performance of oil-producing countries in the face of oil price volatility, which poses a major threat to oil-exporting countries that are heavily dependent on oil revenues.

2.2.3 Oil Dependence Versus Diversification

Natural resource revenues decrease the need for savings and investments in other non-resource-based sectors (Gylfason and Zoega 2002). Countries with resource-based economic development plans may, in the long term, be vulnerable to economic recession and low growth (Payton 2010; Murshed and Serino2011). These economic dilemmas are due mainly to global economic conditions, price fluctuations, ambiguities in reserve estimations and geopolitical uncertainties. Stevens et al. (2015) highlight that high dependence on petroleum revenues jeopardises economic stability and real activity in oil- and gas-abundant countries. They argue that appropriate economic diversification policies have great potential to enable the building of strong, sustainable economies in petroleum-abundant countries. One measure that countries might take in order to combat these economic challenges is to endorse economic and export diversification strategies. In this section, the various factors that might hinder the economic performance of resource-based countries are discussed, with an examination of the different forms of diversification strategy adopted to promote more sustainable economic growth in developing countries with abundant natural resources.

Risks of economic dependence on oil exports

One of the risks of economic dependence on oil exports is the enclave nature of the resources sector, which may hamper the development of the manufacturing sector.

The extractive sector is an enclave because it is capital intensive and has limited links with other sectors of the domestic economy (Gelb 2011; Esanov 2012). In other words, the resources sector generates income independently, without engaging with other domestic economic sectors. It is only domestic labour intensive during the construction phase of the associated infrastructure, so may contribute to high unemployment rates in resource-rich countries (Gelb 2011; Esanov 2012).

As noted above, besides the lack of links with other parts of the economy, resource-dependent countries tend to suffer from poor economic performance because of the crowding out of the traded sector by the extractive sector. In fact, one of the main explanations for Dutch disease is that the traded sector may become less competitive in the international market when squeezed by the natural resources sector (Sachs and Warner 1999). Uncompetitive production by domestic firms may lead to a decline in non-oil export revenues, which in turn increases dependence on the extractive sector in terms of the generation of budget revenue and influence on the exchange rate.

A further factor that may undermine sustainable, long-term economic growth in resource-based economies is the unstable macro-economic situation in many petro-states. Moreover, many oil-producing countries may suffer from macro-economic instability as a result of their economic dependence on oil revenues as the major income-generating channel (Alichi and Arezki 2009).

Implications of diversification

Previous studies have developed a range of arguments associated with the benefits of diversification through the development of the manufacturing sector. In contrast to the extractive sector, manufacturing offers plenty of scope for new job creation, human capital development and innovation. The non-resources sector is also regarded as a catalyst for economic and technological progress, as well as for modernisation. Consequently, diversification is integral to macro-economic stability. Payton (2010) argues that investment in the diversification of economic structure in order to mitigate the negative impact of resource revenue dependence may result in sustainable economic progress. For example, Norway has succeeded in developing a wide range of manufacturing activities, and is therefore less economically dependent on resource revenues. Research suggests that oil-producing countries that suffer from low diversification should learn from the success of other countries in relation to diversification policies and processes. Likewise, Gelb (2011) explains that effective diversification can provide a buffer against oil price shocks, and thus promote macro-economic stability. Sachs and Warner (2001) also insist that development of the manufacturing sector may result in productivity growth and technical advancement.

Another argument for diversification is that it may lead to expansion of employment opportunities in resource-dependent countries. Case studies (Berry 2008) have shown that Indonesia's effective economic policy has resulted in increased job opportunities in various sectors, such as agriculture, manufacturing and construction. Similarly, Chile and Venezuela have focused development on their agricultural sector, which has to some extent helped to create jobs, whereas Nigeria has created more jobs in the construction industry.

Diversification strategies

In order to deal with the negative impacts of over-dependence on the resources sector, it appears imperative to adopt diversification strategies in order to pave the way for sustainable, long-term economic growth. These can be categorised as 'real' and 'quasi-diversification' (Gelb 2011). Real diversification relates to non-resource-based sectors and a move away from the resources sector to manufacturing, agricultural and service sectors, whereas engaging in a resource-based value chain and developing other resource sectors marks quasi-diversification, for example the development of a petrochemical industry based on domestic oil and gas production.

A. Manufacturing sector

The diversification of non-resource-based manufacturing activities may lead to the provision of goods for the local market as import-substituting economic activities, and also for world markets through the promotion of internationally-competitive domestic production. An increase in non-natural resource-based exporting commodities provides resource-rich countries with another source of income that reduces their economic dependence on natural resource revenues (Gelb 2011; Esanov 2012). Many resource-rich countries have shown limited progress in terms of economic diversification; however, Indonesia and Malaysia have effectively diversified their manufacturing sectors. Studies conducted by Gelb (2011) and Esanov (2012) indicate that the measures applied by these two countries have focused on reducing the costs of production, implementing effective macro-economic policy, and upgrading their technological and human capacity.

B. Agricultural sector

Development of the agricultural sector is another form of economic diversification adopted to overcome the resource curse in resource-abundant countries. Hailu et al. (2011) suggest that allocating resource rents to the development of rural areas may result in import substitution in relation to domestic food production and the improvement of food security. An effective agricultural diversification strategy requires public investment in the development of rural infrastructure, irrigation systems, improved seeds and appropriate fertilizers. Coxhead and Li (2008) also argue that effective management of oil rents has enabled Indonesia to diversify its economy into its agricultural sector, which may lead to import-substituting economic activities.

C. Service sector

The development of the service sector, such as tourism and construction industries, may also contribute to the sustainable economic growth of resource-rich countries. For instance, development of the tourism sector is an income source that requires complementary services, such as logistical, technical and health services and well-developed infrastructures, to enable the promotion of diversification, which may also play a major role in reducing the oil revenue dependence of crude oil-exporting countries (Shafaedin 2001).

Value-added economic activities within resources sector

Diversification may also occur within the extractive sector. Industrial sectors provide specialised equipment and services that can be regarded as a backward linkage, while forward linkages may create opportunities to develop resource-based domestic manufacturing activities to generate value-added products such as petrochemicals and refined fuels (Hailu et al. 2011). In this regard, Botswana has succeeded in increasing its manufacturing activities to generate value-added resource-based products by cutting and refining diamonds domestically rather than exporting them uncut for other countries to manufacture the finished product. Therefore, moving the resources sector up the value chain may lead to increased domestic productivity by protecting against technology leakage, thus creating market power (Hailu et al. 2011).

D. Development of other resource sectors

It is widely accepted that the extractive sector may make a significant contribution to the creation of new resource sector activities to promote economic diversification. For example, effective fiscal links between the oil sector and other parts of the economy in Indonesia have enabled development of the gas sector (Gelb 2011).

Impediments to diversification

Diversification of an economy is an immense and challenging task for many resource-dependent countries. The availability of institutional and human capacity may have an impact on the outcome of any diversification strategy. Therefore, well-functioning government institutions and highly skilled human capital may be decisive factors in successful diversification.

Recent studies have shown that an absence of political will and low institutional quality may be major impediments to diversification strategies (Gelb 2011; Esanov 2012; Arezki et al. 2011; Arezki and Nabil 2012). The establishment of strong institutions may also reinforce the transparency and accountability associated with spending of resource revenue by resource-rich countries, as well as providing a more business-friendly climate to attract private domestic and foreign investment. Consequently, leaders have limited room for manoeuvre, because rent seekers often form a major opposition to a diversification policy owing to their desire to maintain their monopoly over the extractive resources of many resource-rich countries (Auty 2001). In other words, economic and political efforts may be severely hampered by leaders whose rent seeking poses a threat to the diversification of manufacturing production and the development of other industrial activities. Other factors that may hinder the implementation of diversification are a lack of skilled human resources and a lack of a well-developed infrastructure.

Lack of access to and the often poor quality of educational systems and the absence of high-impact research and development capacity in extractive countries are further factors that may contribute to low diversification in terms of economic structure and performance (Gylfason 2001). The growth of enterprises and manufacturing requires educated human capital, and policy makers in resource-rich countries may fail to take this aspect into account in their economic policies, so the development of human capacity is likely to be neglected. A study conducted by Ding (2005) and Field shows that resource dependence and human capital development are negatively related. For example, human capacity building has played a major role in Malaysia's successful economic diversification strategy (Gelb 2011).

In summary, natural resources may be a positive driver of economic growth if resource-dependent countries endorse economic diversification and use their revenues to promote sustainable and long-term economic growth. Diversification strategies may take the following forms:

- The improvement of value added within the oil and gas industry by moving down the value chain through the petrochemical industry and the refining sector.
- The development of a domestic oil and gas service industry to capture some of the rent impacts.
- Diversification into other resource sectors such as natural gas and non-fuel minerals.
- Real diversification away from the resources sector into agriculture, manufacturing activities and services such as tourism and construction.
- Export diversification for the purpose of expanding new products with high value added into new export markets.

Skilled human resource capacity and a well-developed infrastructure are integral to success in diversifying economic structure. Review of the literature associated with diversification strategies in resource-dependent countries reveals that institutional quality and political commitment may also be essential factors determining the success of diversification programmes. In other words, diversification may be difficult to achieve when it threatens the monopoly power base of a ruling élite. Thus, successful implementation of the policy measures needed to mitigate economic risks associated with resource dependence rests on the political will to devise the right policies and the institutional capacity to implement them.

2.3 Political Economic Aspects of 'Resource Curse'

According to Ross (1999, p. 307), '[t]he failure of states to take measures that could change resource abundance from a liability to an asset has become the most puzzling part of the resource curse'. Ross highlights that the root cause of poor economic performance in resource-abundant countries is that governments have failed to adopt desired economic policies associated with resource revenue management.

The core of such work is political explanations for ineffective states in terms of combating the 'resource curse'. Theoretical explanations of policy failure have focused on how political incentives generated by large-scale natural resource development may promote rent seeking, corruption and patronage activities by politicians and powerful groups. From this perspective, it is suggested that ineffective governance in relation to the petroleum sector in oil- and natural gas-exporting countries may be related to corruption, rent seeking, patronage, poor institutional quality, lack of skilled human capacity and social and political conflict.

2.3.1 *Governance and Institutions*

The role of government institutions is particularly decisive in the optimal allocation of resource rents in petroleum-exporting countries to productive projects, rather than to 'white elephant' activities that may not be economically efficient but may be of great value to politicians and interest groups (Karl 1999; Torvik 2009; Bridge and Le Billon 2013; Sachs and Stiglitz 2007). In petro-states, government institutions may be either the main contributing factor to a 'resource curse', or a 'blessing'. Therefore, it is essential to reform and develop political and administrative institutions in order to ensure the optimal use of oil and natural gas revenues by securing investment in the productivity of the resources sector and economic diversification.

Similarly, Mehlum et al. (2006) emphasise that the quality of government institutions may determine the impact of resource revenues on the economic performance of resource-abundant countries. They suggest that a 'grabber-friendly institution' may lead to economic stagnation and, in turn, low growth. Similarly, Heredia (1998) highlights the significant role of well-functioning institutions in terms of a sound path towards economic growth in resource-rich countries. However, Sachs and Warner (1995, 1997) do not identify a strong correlation between the economic growth and poor levels of government institutional capacity, measured by rule of law; the reasons for this are discussed in detail in Chap. 3.

2.3.2 *Institutions and Natural Resources*

Arezki and Van der Ploeg (2007) offers a three-fold explanation of the causes of the poor economic performance of some resource-rich countries based on (1) institutions, (2) natural resources and (3) trade policies/openness. Figure 2.2 provides a visual representation of how these three determinants of growth interact with income and with each other.

As can be seen from the directions of relationships in the graph, Arrow 1 indicates that large resource revenues may induce rent-seeking activities, which may result in lower income per capita. Arrows 3 and 4 represent channels through

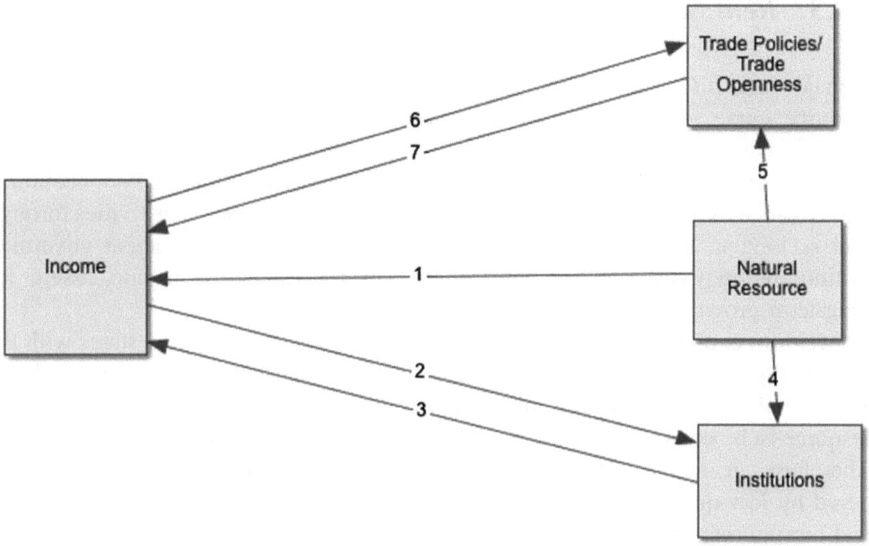

Fig. 2.2 Direct and indirect effects of natural resources on income per capita (Arezkiand Van der Ploeg 2007, p. 12)

which natural resources may aggravate the negative impact of existing low institutional quality on income by increasing corrupt behaviour amongst officials and those in public positions. Arrows 5 and 7 represent the argument that Dutch disease (a decline in non-resource traded commodities due to rises in the real exchange rate) may result in a decline in income per capita. Under this condition, governments are forced by political pressure exerted by interest groups to adopt restrictive trade policies, such as import substitution and subsidies, which reduce income per capita. These inappropriate fiscal policies may accelerate corrupt behaviour, since transfers of oil revenues to protected industries may not be transparent because of low institutional quality. This means that institutional quality may also affect income per capita (Arrows 3 and 7) by diverting resource revenues from more productive use, such as diversification of the wider economy. Arrows 6 and 2 show that institutional quality and trade openness may be affected by income during an oil boom, as huge oil revenues in resource-dependent countries may induce more rent seeking, as well as corruption and patronage.

It can be concluded that the quality of pre-existing institutions may affect the impact of natural resources on growth while, conversely, resource abundance may also have an effect on institutional quality, leading to ineffective governance through rent seeking, corruption and patronage.

2.3.3 *Rent Seeking*

A major focus of research associated with the resource curse is the link between the resource curse and rent-seeking behaviour. Mehlum et al. (2006) explain that entrepreneurs tend to specialise in rent-seeking rather than productive activities due to high resource revenues and the poor quality of public institutions. Poor economic performance may be attributable to the escalation of unproductive activities through rent seeking in oil- and gas-rich countries. In other words, inefficient governing institutions may waste resources rather than translating them into assets for long-term growth.

Mehlum et al. (2006) also link high growth in resource-abundant states with the effective rule of law, which may enhance institutional quality. It is suggested that well-functioning institutions may account for economic progress in some resource-rich countries, such as Norway, Botswana, Canada and Australia. Therefore, an increase in productivity, and thus economic growth, may be hampered by low-quality institutions, creating the potential for increased rent seeking and corruption among the powerful.

Kolstad (2009) argues that increased rent seeking and a tendency towards patronage are aggravated by inefficient governing institutions, and highlights that the development of institutional quality may contribute to the sound management of natural resources. This means that an effective state may turn its resource endowment into a blessing rather than a curse. The same author mentions that rent-seeking incentives and patronage tend to emerge among powerful groups and politicians, which may cause a decline in the productivity of public sector activities and projects. Several other authors also argue that natural resources may impede democratisation by increasing rent-seeking and patronage incentives in countries excessively dependent on oil and gas revenues, namely rentier states (Ross 2001; Busse and Gröning 2013).

2.3.4 *Corruption*

The literature dealing with the resource curse (e.g. Karl 2004) explains corrupt behaviour as a tendency by politicians and officials to satisfy individual interests through the illegitimate use of authority, while sacrificing the public interest associated with the distribution of resource rents. Corruption is a consequence of poor institutions and may harm a resource-abundant state's efforts to enhance its economic performance. For instance, Al-Kasim et al. (2013) argue that weak petroleum governance may have a negative impact on the provision of welfare in oil-exporting countries. In this regard, corruption between those in authority and oil firms is considered to be a major obstacle to optimal oil and gas production. Under such circumstances, corruption may occur in the form of bribes, cash or payments in kind by oil firms to officials to secure access to the national reserve base.

In this setting, another study also considers the state's involvement in petroleum production in oil-producing countries through national oil companies to be a determining factor for increased corruption among officials related to the oil and gas sector. For instance, Arezki and Bruckner (2011) highlight that the ownership structure of oil and petroleum industries plays a major role in the existence of corruption in the resources sector. Furthermore, rentier states are granted greater authority over resource development, and thus oil rents directly accruing to the government budget may create more opportunities for corrupt state leaders in connection with oil production because of poor institutional practices and a lack of transparency. Consequently, rentier states' attempts to reduce political rights in order to promote their own interests are the most likely reasons for conflict and civil war (Ross 2001).

Similarly, Hammond's (2011) case studies on Angola and Venezuela indicate that increased political control over oil companies and state participation in petroleum production have encouraged corruption. Nigeria is another frequently-cited example of a corrupt country associated with oil revenues (Osoba 1996; Watts 2004). Widespread corruption in Nigeria has resulted in inefficiency in terms of domestic investment and low-quality projects, both of which may impact negatively on economic development.

2.3.5 Patronage

In addition to corruption and rent-seeking behaviour, the existence of a patronage system among politicians and other powerful groups is another reason for poor institutional capacity in resource-dependent countries. It is widely accepted that in many resource-rich countries, where abuse and mismanagement of resource rents has resulted in low growth, governments may try to remain in power through the distribution of resource rents to those in official positions and other powerful groups (Kolstad and Soreide 2009).

To counter the problems discussed above, Collier and Hoeffler (2005) propose a model of democratic politics that compromises between electoral competition and checks and balances to impede patronage behaviours. A lack of effective functioning institutions may result in poor accountability and transparency in resource revenue allocation and spending. Under such circumstances, governments tend to use resource rents for their own patronage purposes, which may bring a democratic regime into question. Similarly, Robinson et al. (2006) emphasise that resource rents tend to be used by politicians for patronage purposes in order to generate political support and achieve re-election. For example, the expansion of public spending was a contributory factor in the results of parliamentary and presidential elections in Russia in 2007 (Sturm et al. 2009). Another way for political leaders to increase patronage is the allocation of lucrative and influential public sector posts to their supporters.

2.3.6 Other Political Economic Explanations

Inequality and social conflict are other negative consequences of poor institutional quality in terms of the redistribution of resource rents in extractive countries. Ramsay (2011) argues that there may be a negative correlation between increasing petroleum revenue and institutional capacity in oil-producing countries. This negative relationship, which may result in inequality and authoritarianism, arises from such factors as the diverse economic and political histories of resource-rich countries. According to Johan (2007), political instability may result in poor economic performance in oil-producing countries. Social and political conflict may be caused by a lack of transparency and accountability in terms of revenue management in resource-rich countries. In this context, rent seekers attempt to increase their profits from oil revenues, which in turn generates inequality in the distribution of oil income, and is likely to cause social conflict and even civil war in resource-rich countries.

Democracy in extractive countries is perhaps under most threat from strong competition between powerful groups to gain profit from resource revenues. The lack of effective systems of checks and balances in the resources sector, linked to weak fiscal institutions, may lead to the mismanagement of oil revenue collection and spending. Elbadawi and Soto (2012) and Collier (2010) argue that many resource-rich countries have been afflicted by the resource curse due to poor monitoring mechanisms in relation to the flow of rents from the exploitation of natural resources. They argue that this weakness on the part of institutions is inconsistent with the practice of democracy in oil-exporting states.

As discussed above, the development of human resource capacity is often ignored in petroleum-exporting countries in terms of failing to fund education adequately (Gylfson 2001; Papyrakis and Gerlagh 2004; Humphreys et al. 2007). Furthermore, human resource development is delayed by a lack of resource revenue allocation to industries with intensive learning by doing (Van Wijnbergen 1984). As a result, many such countries may suffer from poor institutional quality. Kronenberg (2004) explains that inefficient institutions may stem from unprepared and unskilled actors in relation to the resources sector. Under these conditions, a rise in corruption and an ineffective rule of law endanger the public interest.

Political and economic instability, which may seriously damage a resource-rich country's economy, may be due to the mismanagement of oil rents through poor institutional capacity. Guenther (2008) stresses that a resource curse may develop in resource-dependent countries as a result of a weak political economy and bad governance associated with the resources sector. Ruling élites tend to gain profits by increasing their control over the resources sector. In this context, rentier states levy low taxes because a large proportion of oil revenues accrue directly to these states. This leads to a decline in the need for an effective taxation system (Ross 1999; Karl 1999; Mahdavy 1970; Collier and Hoeffler 2005), and consequently these governments do not meet their commitment to be held accountable for the spending of resource revenues in economies over-dependent on these revenues.

2.3.7 Accountability and Transparency

Many authors dealing with the resource curse attribute its root cause to a lack of institutional quality in resource-rich countries such as petroleum exporters. However, producer-friendly institutions may combat the emergence of the resource curse and promote the creation of wealth within resource-rich nations (Mehlum et al. 2006). These producer-friendly institutions seek transparency and accountability, which may provide substantial opportunities to increase productive activities and so encourage long-term economic growth. Kolstad and Wiig (2008) assert that transparency is a key factor in curbing the negative impacts of the resource curse. Improvement of institutional quality through transparency, making information with respect to resource revenue collection and spending available to public scrutiny, may reduce corruption and patronage. These authors point out that the emergence of various international initiatives may enforce extractive states to improve their institutional quality, thereby promoting transparency. High-quality institutions may encourage human resource development, which may be regarded as a pillar of economic growth. Butkiewicz and Yanikkaya (2010) indicate that producer-friendly institutions may enhance human resource capacity in resource-rich countries.

The implementation of transparency and accountability may result in an increase in public trust, which may resolve political and social tensions in oil- and natural gas-exporting countries. In this regard, Herringshaw (2004) argues that the poor economic performance of resource-abundant countries may be attributed to a lack of transparency in resource revenue management. For instance, the governments of many African natural resource extractive countries are unable to handle resource income effectively in order to improve the welfare of their nations, due mainly to poor institutional quality. In such contexts, social conflict is highly likely.

2.4 Conclusions

A key shortcoming of the resource curse literature is the lack of clear understanding of the effect of the managerial model of the oil sector, in particular its effects on the sector's economic performance. Few case studies have focused on effective administrative design as a causal factor affecting the performance of the oil sector. The literature suggests that an effective model of governance for the national petroleum sector enables the effective management of petroleum resources and revenues (Heller and Marcel 2012; Lahn et al. 2009; Thurber et al. 2010a, b; 2011; Luong and Weinthal 2006).

The main contribution of the current research will be to identify the main channels through, which oil and gas resources may promote or impede economic growth from a broader social perspective. This information will be used to outline the main policy implications for domestic policy makers in achieving desirable growth outcomes. In other words, this research will provide a social explanation for the resource curse in resource-rich countries. Therefore, high priority should be given to policies that address the enhancement of 'social capital'. This means that policies in resource-abundant countries should be less about macro-economic management and more about administrative and human capacity building to reduce or prevent the negative impacts of the resource curse, thus providing the right incentives to players in the resources sector. Moreover, all policies need to take into account existing bureaucratic capacity—the government's capability to deliver effective policies and public services independently of political pressures and interventions—and human resource capacity as determining factors in success or failure.

References

Al-Ezzee I (2011) Real influences of real exchange rate and oil price changes on the growth of real GDP: case of Bahrain. Int Proc Econ Dev Res 8:155–164

Alichi A, Arezki R (2009) An alternative explanation for the resource curse: the income effect channel. IMF. Working Paper 09/112, International Monetary Fund, Washington, DC

Al-Kasim F, Søreide T, Williams A (2013) Corruption and reduced oil production: an additional resource curse factor? Energy Policy 54:137–147

Al-Marhubi F (1998) Cross-country evidence on the link between inflation volatility and growth. Appl Econ 30:1317–1326

Arezki R, Blanchard O (2015) The 2014 oil price slump: seven key questions. VOX, 13 January. Available at: http://www.voxeu.org/article/2014-oil-price-slump-seven-key-questions. Accessed 12 Feb

Arezki R, Bruckner M (2011) Oil rents, corruption, and state stability: evidence from panel data regressions. Eur Econ Rev 55:955–963

Arezki R, Nabil MK (2012) Natural resources, volatility, and inclusive growth: perspectives from the Middle East and North Africa. CESifo. Working Paper No. 3818, Munich, Germany

Arezki R, Van der Ploeg F (2007) Can the natural resource curse be turned into a blessing? The role of trade policies and institutions. IMF Working Paper 07/55, International Monetary Fund, Washington, DC

Arezki R, Gylfason T, Sy A (2011) Beyond the curse: policies to harness the power of natural resources. International Monetary Fund, Washington, DC

Auty RM (1993) Sustaining development in mineral economies: the resource curse book. Routledge, London

Auty RM (ed) (2001) Resource abundance and economic development. Oxford University Press, Oxford

Bacon R, Kojima M (2008) Coping with oil price volatility: energy sector management assistance program report. The World Bank, Washington, DC

Baffes J, Kose MA, Ohnsorge F, Stocker M (2015) The great plunge in oil prices: causes, consequences, and policy responses. World Bank Group Policy Research Note PRN/15/01, The World Bank, Washington, DC

Barrell R, Kirby S (2008) Oil prices and growth. National Inst Econ Rev 204(1):39–42

Benedictow A, Fjaertoft D, Løfsnaes O (2013) Oil dependency of the Russian economy: an econometric analysis. Econ Modell 32:400–428

Berry A (2008) Growth, employment and distribution impacts of minerals dependency: four case studies. S Afr J Econ 76(S2):S148–S174

BP (2015) BP statistical review of world energy, 64th edn. Available at: http://www.bp.com/content/dam/bp/pdf/energy-economics/statistical-review-2015/bp-statistical-review-of-world-energy-2015-full-report.pdf. Accessed 11 Feb 2016

Bridge G, Le Billon P (2013) Oil. Polity Press, Cambridge

Bunte J (2011) Why does the severity of the Dutch disease vary across countries? Working Paper, Department of Political Science, University of Minnesota, Minneapolis, MN

Busse M, Gröning S (2013) The resource curse revisited: governance and natural resources. Public Choice 154(1–2):1–20

Butkiewicz JL, Yanikkaya H (2010) Minerals, institutions, openness, and growth: an empirical analysis. Land Econ 86(2):313–328

Collier P (2010) The political economy of natural resources. Soc Res 77(4):1105–1132

Collier P, Hoeffler A (2005) Democracy and resource rents. Mimeograph, University of Oxford, Oxford

Cologni A, Manera M (2013) Exogenous oil shocks, fiscal policies and sector reallocations in oil producing countries. Energy Econ 35:42–57

Cordon WM, Neary JP (1982) Booming sector and deindustrialisation in a small open economy. Econ J 92(368):825–848

Coutinho L (2011) The resource curse and fiscal policy. Cyprus Econ Policy Rev 5(1):43–70

Coxhead I, Li M (2008) Prospects for skills-based export growth in a labour-abundant, resource-rich developing economy. Bull Indones Econ Stud 44(2):209–238

Davis GA, Tilton JE (2005) The resource curse. Nat Resour Forum 29(3):233–242

Davis J, Ossowski R, Daniel JA, Barnett S (2003) Stabilisation and savings funds for non-renewable resources: experience and fiscal policy implications. In: Davis JM, Ossowski R, Fidelino A (eds) Fiscal policy formulation and implementation in oil-producing countries. International Monetary Fund, Washington, DC, pp 273–315

Dées S, Gasteuil A, Kaufmann RK, Mann M (2008) Assessing the factors behind oil price changes. Working Paper 855, European Central Bank, Frankfurt, Germany

Di John J (2011) Is there really a resource curse? A critical survey of theory and evidence. Glob Gov 17(2):167–184

Ding N, Field BC (2005) Natural resource abundance and economic growths. Land Econ 81 (4):496–502

Dobrynskaya V, Turkisch E (2010) Economic diversification and Dutch disease in Russia. Post-Communist Econ 22(3):283–302

Dudley D (2011) Political risk for oil producer. Middle East Econ Digest 55(10):26–27

El Anshasy AA, Bradley MD (2012) Oil prices and the fiscal policy response in oil-exporting countries. J Policy Model 34(5):605–620

Elbadawi I, Soto R (2012) Resource rents, political institutions and economic growth. Working Paper 678, Economic Research Forum, Giza, Egypt

Esanov A (2012) Economic diversification: dynamics, determinants and policy implications. Revenue Watch Inst 8

Fouche G, Milhench C (2015) After glory days, cheap oil forces sovereign funds to retreat. The Reuters, 7 December. Available at: http://www.ft.com/cms/s/0/ebe4acc0-1cef-11e1-a134-00144feabdc0.html# axzz3zw63544a. Accessed 18 Mar 2016

Gelb A (2011) Economic diversification in resource rich countries. Mimeograph, Center for Global Development, Washington, DC

Guenther B (2008) The Asian drivers and the resource curse in Sub-Saharan Africa: the potential impacts of rising commodity prices for conflict and governance in the DRC. Eur J Dev Res 20 (2):347–363

Gylfason T (2001) Natural resources, education, and economic development. Eur Econ Rev 45 (4–6):847–859

Gylfason T, Zoega G (2002) Inequality and economic growth: do natural resources matter? CESifo. Working Paper No. 712, Munich, Germany

Hailu D, Rendtorff-Smith S, Gankhuyag U, Ochieng C (2011) Conflict prevention in resource-rich economies. United Nations Development Programme, New York. Available at: http://www.un.

org/en/land-natural-resources-conflict/pdfs/Resource%20Rich%20Economies.pdf. Accessed 12 Feb 2016
Hammond JL (2011) The resource curse and oil revenues in Angola and Venezuela. Sci Soc 75 (3):348–378
Heller PRP, Marcel V (2012) Institutional design in low-capacity oil hotspots. Revenue Watch Institute, New York
Heredia B (1998) Review of Terry Lynn Karl 'The paradox of plenty: oil booms and petro-states'. Can J Polit Sci 31, 815–16
Herringshaw V (2004) Natural resources—curse or blessing? New Econ 11(3):174–177
Humphreys M, Sachs JD, Stiglitz JE (2007) Introduction: what is the problem with natural resource wealth? In: Humphreys M, Sachs JD, Stiglitz JE (eds) Escaping the resource curse. Columbia University Press, New York, pp 1–20
Ickes BW, Gaddy CG (2005) The virtual economy revisited: resource rents and the Russian economy. In: Keynote speech delivered at renaissance capital's 4th annual equity conference, 'Russia: New Investment Horizons', The Metropolitan Club, New York City, NY, 20 Oct
Ismail K (2010) The structural manifestation of the Dutch disease: the case of oil exporting countries. IMF Working Paper no. 10/103, International Monetary Fund, Washington, DC
Johan JD (2007) Oil abundance and violent political conflict: a critical assessment. J Dev Stud 43 (6):961–986
Karl TL (1999) The perils of the petro-state: reflections on the paradox of plenty. J Int Aff 53 (1):31–38
Karl TL (2004) Oil-led development: social, political, and economic consequences. In: Cleveland CJ (ed) Encyclopaedia of energy. Elsevier, Amsterdam, The Netherlands, pp 661–672
King K, Deng A, Metz D (2012) An econometric analysis of oil price movements: the role of political events and economic news, financial trading, and market fundamentals. Bates White Economic Consulting, Washington, DC
Kolstad I, Søreide T (2009) Corruption in natural resource management: implications for policy makers. Resour Policy 34(4):214–226
Kolstad I, Wiig A (2008) Political economy models of the resource curse: implications for policy and research. CMI Working Paper WP2008:6, Chr. Michelsen Institute, Bergen, Norway
Kronenberg T (2004) The curse of natural resources in the transition economies. Econ Transit 12 (3):399–426
Kuralbayeva K, Stefanski RR (2013) Windfalls, structural transformation and specialization. OxCarre Research Paper 54, Oxford Centre for the Analysis of Resource Rich Economies, Oxford
Labban M (2010) Oil in parallax: scarcity, markets, and the financialisation of accumulation. Geoforum 41:541–552
Lahn G, Marcel V, Mitchell J, Myers K, Stevens P (2009) Good Governance of the National Petroleum Sector: the Chatham House document. The Royal Institute of International Affairs, London
Lipscomb L, Ramakrishnan U, Adler G, Bundina N, Xiangming L (2010) International monetary fund Indonesia. International Monetary Fund, Washington, DC
Lombardi M, Van Robays I (2011) Do financial investors destabilise oil prices? Working Paper No. 1346, European Central Bank, Frankfurt, Germany
Lorde T, Jackman M, Thomas C (2009) The macroeconomic effects of oil price fluctuations on a small open oil-producing country: the case of Trinidad and Tobago. Energy Policy 37(7):2708–2716
Luong PJ, Weinthal E (2006) Rethinking the resource curse: ownership structure, institutional capacity, and domestic constraints. Annu Rev Polit Sci 9:241–263
Mahdavy H (1970) The patterns and problems of economic development in rentier states: the case of Iran. In: Cook MA (ed) Studies in the economic history of the Middle-East. Oxford University Press, Oxford, pp 428–467

Medas P, Zakharova D (2009) A primer on fiscal analysis in oil-producing countries. IMF Working Paper 09/56, International Monetary Fund, Washington, DC
Mehlum H, Moene K, Torvik R (2006) Institutions and the resource curse. Econ J 116(508):1–20
Murshed SM, Serino LA (2011) The pattern of specialization and economic growth: the resource curse hypothesis revisited. Struct Change Econ Dyn 22(2):151–161
Osoba SO (1996) Corruption in Nigeria: Historical perspective. Rev Afr Polit Econ 23(69):371–386
Ossowski R, Villafuerte M, Medas PA, Thomas T (2008) Managing the oil revenues boom: the role of fiscal institutions. International Monetary Fund, Washington, DC
Papyrakis E, Gerlagh R (2004) The resource curse hypothesis and its transmission channels. J Comp Econ 32(1):181–193
Payton S (2010) Without portfolio. Financial Management, June, 22–26
Ramey G, Ramey VA (1995) Cross-country evidence on the link between volatility and growth. Am Econ Rev 85(5):1138–1151
Ramsay KW (2011) Revisiting the resource curse: natural disasters, the price of oil, and democracy. Int Organ 65:507–529
Robinson JA, Torvik R, Verdier T (2006) Political foundations of the resource curse. J Dev Econ 79(2):447–468
Ross ML (1999) The political economy of the resource curse. World Polit 51(2):297–322
Ross ML (2001) Does oil hinder democracy? World Polit 53(3):325–361
Ross ML (2012) The oil curse: how petroleum wealth shapes the development of nations. Princeton University Press, London
Saad-Filho A, Weeks J (2013) Curses, diseases and other resource confusions. Third World Q 34 (1):1–21
Sachs JD, Stiglitz JE (2007) Escaping the resource curse. Columbia University Press, New York
Sachs DJ, Warner AM (1995) Natural resource abundance and resource growth. Working Paper 5398, National Bureau of Economic Research, Cambridge, MA. Available at: http://www.nber.org/papers/w5398. Accessed 12 Feb 2016
Sachs DJ, Warner AM (1997) Natural resource abundance and economic growth. Research Paper, Centre for International Development and Harvard Institute for International Development, Cambridge, MA
Sachs J, Warner AM (1999) The big push, natural resource booms and growth. J Dev Econ 59:43–76
Sachs JD, Warner AM (2001) The curse of natural resources. Eur Econ Rev 45(4):827–838
Shafaeddin M (2001) Diversification, employment and development: Towards a long-term strategy for an oil-exporting country—the case of Iran. Working Paper 0113, Economic Research Forum, Giza, Egypt
Sornette D, Woodard R, Zhou W (2009) The 2006-2008 oil bubble: evidence of speculation, and prediction. Phys A 388:1571–1576
Stevens P, Lahn G, Kooroshy J (2015) The resource curse revisited. Research Paper, Chatham House, Royal Institute of International Affairs, London
Sturm M, Gurtner F, Alegre GJ (2009) Fiscal policy challenges in oil-exporting countries: a review of key issues. ECB Occasional Paper 104, European Central Bank, Frankfurt, Germany
Tabata S (2012) Observations on Russian exposure to the Dutch disease. Eurasian Geogr Econ 53 (2):231–243
Thurber MC, Emelife IM, Heller PR (2010a) NNPC and Nigeria's oil patronage ecosystem. Working Paper 95, Program on Energy and Sustainable Development, Stanford University, Stanford, CA
Thurber M, Hults D, Heller PR (2010b) The limits of institutional design in oil sector governance: exporting the 'Norwegian Model'. Paper presented at the international studies association annual convention, New Orleans, LA, 18 February
Thurber MC, Hults DR, Heller PR (2011) Exporting the 'Norwegian Model': the effect of administrative design on oil sector performance. Energy Policy 39(9):5366–5378

Torvik R (2009) Why do some resource-abundant countries succeed while others do not? Oxford Rev Econ Policy 25(2):241–256
Usui N (2007). How effective are oil funds? Managing resource windfalls in Azerbaijan and Kazakhstan. ERD Policy Brief Series No. 50, Asian Development Bank, Manila, Philippines
Van der Ploeg F, Poelhekke S (2009) Volatility and the natural resource curse. Oxford Econ Pap 61(4):727–760
Van Wijnbergen S (1984) 'The Dutch disease': a disease after all? Econ J 94(373):41–55
Van Wijnbergen S, Budina N (2011) Fiscal sustainability, volatility and oil wealth. Econ Transit 19(4):639–666
Watts M (2004) Resource curse governmentality, oil and power in the Niger Delta, Nigeria. Geopolitics 9(1):50–80
Zhang YJ, Fan Y, Tsai HT, Wei YM (2008) Spillover effect of US dollar exchange rate on oil prices. J Policy Model 30(6):973–979

Chapter 3
Transmission Channels of the 'Resource Curse' Reappraised

Abstract This chapter reappraises the negative relationship between natural resources and economic growth by applying empirical cross-sectional and panel data analysis. The empirical results show the importance of economic, political and social transmission mechanisms of the resource curse with regard to the different types of natural resource, particularly in fuel-exporting countries. The results of both cross-sectional and panel data models indicate that high economic dependence on fuel exports leads to lower economic performance. Main findings show that the effect of institutions is lower in highly fuel-dependent countries than in those with economies less reliant on fuel export revenues. Moreover, the economic performance of fuel-dependent economies is more likely to suffer from the detrimental effects of the resource curse.

Keywords Resource curse · Cross-sectional analysis · Panel data analysis · Transmission channels of resource curse · Fuel dependent economies · Institutional quality

3.1 Introduction

The objectives of this chapter are to investigate whether there is evidence of the resource curse—a significant negative relationship between natural resource dependence and economic growth—and, if so, to reappraise the importance of transmission channels with regard to the resource curse.

The main research questions are:

- is there a negative connection between resource dependence and economic growth?
- does human resource and institutional development positively affect growth performance, and do natural resources influence growth through institutional and human resource quality?

K. Abdullah Auzer, *Institutional Design and Capacity to Enhance Effective Governance of Oil and Gas Wealth: The Case of Kurdistan Region*, Perspectives on Development in the Middle East and North Africa (MENA) Region, DOI 10.1007/978-981-10-4518-9_3

This research reappraises the negative relationship between natural resources and economic growth by applying empirical cross-sectional and panel data analysis using a large dataset of 160 countries for the period 1970–2010. In these analyses, the emphasis is on (1) the applicability of economic variables to economic growth, and (2) the specific influence of institutional and human resource capacity on natural resources in natural resource-dependent economies.

This chapter proceeds as follows. Section 3.2 presents the samples and a description of the data. Section 3.3 introduces the generic growth model and provides an overview of the most common methodological difficulties encountered in empirical studies seeking to confirm resource curse effects. Section 3.4 applies a cross-sectional model and Sect. 3.5 uses panel data with a five-year average to estimate a dynamic panel growth regression model. Section 3.6 presents a summary and conclusions.

3.2 Econometric Model

The basic econometric specification for testing whether natural resources induce certain distortions in the economy resulting in low economic growth is a growth regression model using elements from neoclassical and endogenous growth theories, as described by Durlauf et al (2005). This model postulates a conditional convergence hypothesis, which states that economic growth is negatively related to initial income (Solow 1994 ; Lee et al. 1997). The growth equation has generally been presented as follows:

$$g_{it} = \beta_0 + \beta_i g_{it-1} + \beta_2 X_{it} + \beta_3 NR_{it} + \beta_4 Inst_{it} + \beta_5 HR_{it} + \varepsilon_{it} \quad (3.2.1)$$

where g is the growth rate in per worker GDP over the period, g_{t-1} is the log of per worker t,i at time t in country i, X_i is a vector of economic variables, and NR is an indicator of resource dependence. $Inst$ and HR are measures of institutional and human resource quality, respectively. In estimating the equation, the most common methodological difficulties for empirical studies are: (1) the choice of control variables, (2) the choice of adequate estimators, (3) the measure of the natural resource, and (4) the effect of different types of natural resource dependence on GDP growth. The lagged value of GDP per worker accounts for the relative convergence hypothesis.

3.3 Sample and Data Availability

The sample and data used in this analysis consist of all countries and years for which relevant data are available, covering 160 countries from 1970 to 2010. The initial year of the dataset is 1970 because, on average, the major resource exporters had high incomes at the beginning of this period as oil prices began to increase in the early 1970s. Sachs and Warner (1995) argue that natural resource-rich countries with high incomes in 1970 tended to exhibit lower growth over the following forty years.

The data used to construct the regressions were drawn from two databases frequently used in the literature: the World Bank World Development Indicators (WDI) and the Penn World Tables. GDP per worker, measured by purchasing power parity (PPP), is expressed in constant 2005 international dollars. Prior to taking the natural log, investment as a share of GDP was obtained from Penn World Table 3.7. Average years of secondary schooling for the entire population, terms of trade, and data on primary exports for the period 1970–2010, and institutional variables were drawn from the World Bank WDI Governance Database.

Descriptive statistics tables provide relevant information, including the number of available observations and the general direction of responses. For example, Table 3.1 indicates that the mean share of fuel exports as a proportion of GDP was higher than the mean share of mineral exports and agriculture exports for the entire sample in the cross-sectional data. It also indicates that between 1970 and 2010 there was greater variance in fuel exports than in other types of natural resource.

In order to shed light on time variations in the dataset, descriptive statistics for two sub-periods are presented: before 1995, and after 1995 when oil prices were rising steadily (see Tables 3.2 and 3.3). As can be seen, the fuel export variable shows less variability in the years after 1995. However, there is a minor difference

Table 3.1 Characteristics of sample data (n = 160 observations) for cross-sectional model

Variable	Abbreviation	Obs	Mean	Std. dev.	Min	Max
Natural resource exports	NrEXP					
Fuel export/GDP %	FuelEXP_sgdp	156	6.44	13.47	0	73.68
Mineral export/GDP %	MinExp_sgdp	156	2.67	6.22	0	46.92
Agriculture export/GDP %	AgrExp_sgdp	156	1.33	2.16	0	14.12
Human resources	HR	157	56.46	31.21	4.80	139.62
Institution/accountability	EstVA	159	−0.06	0.96	−1.90	1.61
Institution/political stability	EstPS	159	−0.11	0.97	−2.81	1.51
Institution/government effectiveness	EstGEF	159	0.01	1.00	−2.18	2.15
Institution/regulatory quality	EstRQ	159	0	0.96	−2.44	1.92
Institution/rule of law	EstRL	159	−0.06	1.01	−2.35	1.94
Institution/control of corruption	EstCC	159	0	1.01	−1.74	2.45
Investment/GDP %	Inv	159	22.99	8.25	3.73	44.99
Terms of trade growth rate 2000–2010	TOTG	157	1.24	3.30	−12.25	8.57
Initial income	LogGDP70	156	9.15	1.31	6.59	12.28
GDP growth rate 1970–2010	Growth70–10	156	1.05	1.72	−4.01	7.20

Table 3.2 Descriptive statistics for period before 1995 for panel data model

Variable	Abbreviation	Obs	Mean	Std. dev.	Min	Max
Natural resource exports	NrExp					
Fuel export/GDP %	FuelExp_Sgdp	421	5.62	13.27	0.00	93.52
Mineral export/GDP %	MinExp_Sgdp	436	2.28	6.28	0.00	57.08
Agricultural export/GDP %	AgerExp_Sgdp	438	1.38	2.33	0.00	16.66
Human resource	HR	552	50.70	31.47	1.90	132.86
Institution/accountability	EstAV	0				
Institution/political stability	EstPS	0				
Institution/government effectiveness	EstGEF	0				
Institution/regulatory quality	EstRQ	0				
Institution/rule of law	EstRL	0				
Institution/control of corruption	EstCC	0				
Terms of trade	TOT	242	117.66	44.21	35.61	320.94
Investment/GDP %	Inv	615	22.97	11.02	1.46	63.02
GDP growth rate	GDP Growth	617	0.90	3.33	−19.84	12.02

Table 3.3 Descriptive statistics for period after 1995 for panel data model

Variable	Abbreviation	Obs	Mean	Std. dev.	Min	Max
Natural resource exports	NrExp					
Fuel export/GDP %	FuelExp_Sgdp	537	6.03	12.14	0.00	64.16
Mineral export/GDP %	MinExp_Sgdp	548	2.23	4.72	0.00	38.23
Agricultural export/GDP %	AgerExp_Sgdp	546	1.03	2.01	0.00	18.16
Human resource	HR	522	71.87	32.04	5.38	154.05
Institution/accountability	EstAV	624	−0.06	0.97	−2.07	1.66
Institution/political stability	EstPS	623	−0.12	1.00	−3.11	1.64
Institution/government effectiveness	EstGEF	624	0.00	1.01	−2.31	2.26
Institution/regulatory quality	EstRQ	624	0.00	0.98	−2.56	2.23
Institution/rule of law	EstRL	624	−0.06	1.02	−2.51	1.98
Institution/control of corruption	EstCC	624	0.00	1.03	−1.98	2.52
Terms of trade	TOT	545	109.13	29.81	24.99	235.61
Investment/GDP %	Inv	625	23.32	9.57	2.59	76.51
GDP growth rate	GDP Growth	625	0.81	3.61	−22.74	36.52

between the two sub-periods in the standard deviation of the GDP growth rate, the maximum GDP growth being higher over the latter period. Furthermore, Table 3.2 has missing observations for institutional indices in the years before 1995.

Table 3.4 presents a correlation matrix for the main variables. Note that the correlation is negative between all types of natural resource and the economic growth rate, but the relationship is stronger with fuel exports than with mineral and agricultural exports. Interestingly, the share of fuel exports in GDP is more positively correlated with initial income than the negative and weaker relationship between mineral and agricultural exports and initial income. Furthermore, a positive link is found between investment, openness and terms of trade and both fuel and mineral exports' share of GDP.

Table 3.4 also indicates that the fuel export share is positively associated with human resources but negatively correlated with quality of institution. Similarly, shares of agricultural and mineral exports are negatively correlated with institutional quality. However, quality of institution has quite a strong positive association with the human resource variable, exhibiting the strongest relationship between the different variables in the correlation matrix. Moreover, collinearity might be an issue when looking at high correlation among measures of institutional quality, however the results revealed that some of them are actually significant when they are included.

Figures 3.1, 3.2 and 3.3 present scatter plots of the relationships between growth in real GDP per worker from 1970 to 2010, and primary exports (fuel, minerals and agriculture) as a share of GDP. These plots indicate that the natural resource intensity of exports is negatively related to economic growth, particularly in countries that have larger fuel and mineral exports. Furthermore, A regression specification that allows for different effects of fuel exports on growth depending on the size of the share of fuel exports to GDP is considered in Sect. 3.4.2 by employing[1] dummy variables.

3.3.1 *Control Variables*

The first decision to be made in modelling research associated with the resource curse is the choice of potential growth determinants as control variables. As explained earlier, the resource curse literature has used a wide range of factors to account for the economic performance of resource-rich countries. The following control variables were selected on the basis of economic transmission channels of the resource curse described in the literature:

---

[1]Another way is to apply a generalized Tobit (Heckman two-step) or count data model with zero-inflation. In the two-step procedure, first one estimates a Probit model of the determinant of natural resources and then in the second step the degree of dependence/growth effect to correct for selection bias. The use of a count data model accounts for the large number of zeros in the data.

Table 3.4 Correlation matrix

	Growth 70–10	Log GDP 70	FuelEXP _sgdp	MinExp _sgdp	AgrExp _sgdp	HR	EstVA	EstPS	EstGEF	EstRQ	EstRL	EstCC	Inv	TOTG
Growth 70–10	1													
Log GDP 70	−0.24	1												
Natural resource exports														
FuelEXP_sgdp	−0.23	0.4	1											
MinExp_sgdp	−0.28	−0.05	−0.07	1										
AgrExp_sgdp	−0.05	−0.27	−0.12	0.34	1									
HR	0.15	0.77	0.11	−0.13	−0.24	1								
EstVA	0.27	0.43	0.3	−0.1	−0.09	0.63	1							
EstPS	0.26	0.49	−0.01	−0.08	−0.09	0.56	0.69	1						
EstGEF	0.34	0.6	−0.07	−0.13	−0.21	0.75	0.81	0.75	1					
EstRQ	0.32	0.57	−0.08	−0.11	−0.2	0.71	0.83	0.71	0.95	1				
EstRL	0.34	0.57	−0.07	−0.13	−0.18	0.73	0.83	0.81	0.96	0.93	1			
EstCC	0.26	0.59	−0.07	−0.1	−0.16	0.72	0.8	0.78	0.96	0.9	0.96	1		
Inv	0.29	0.22	0.04	0.02	−0.07	0.26	0.17	0.28	0.29	0.29	0.31	0.24	1	
TOTG	−0.23	0.13	0.45	0.23	0.11	−0.08	−0.29	−0.15	−0.18	−0.21	−0.2	−0.16	−0.05	1

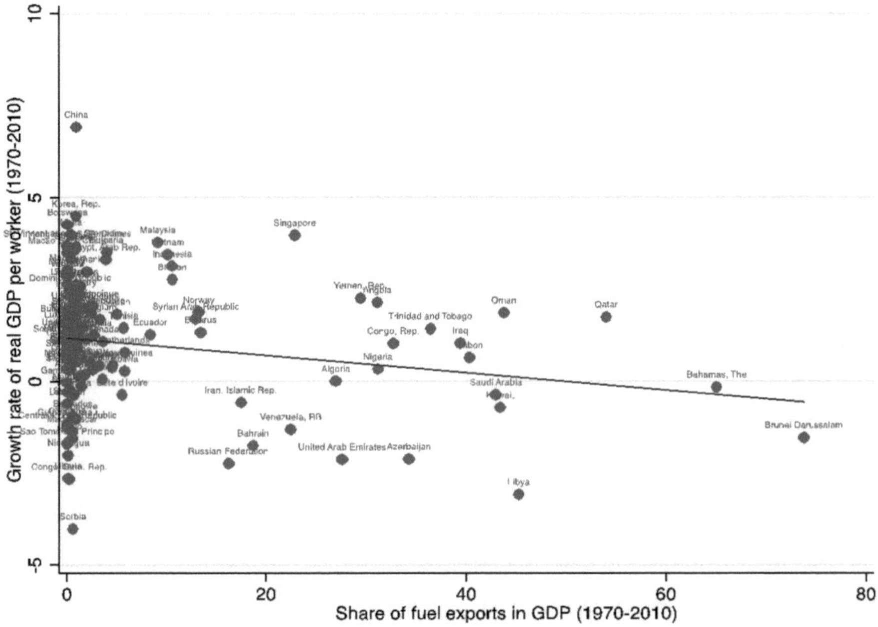

Fig. 3.1 Fuel exports and economic growth

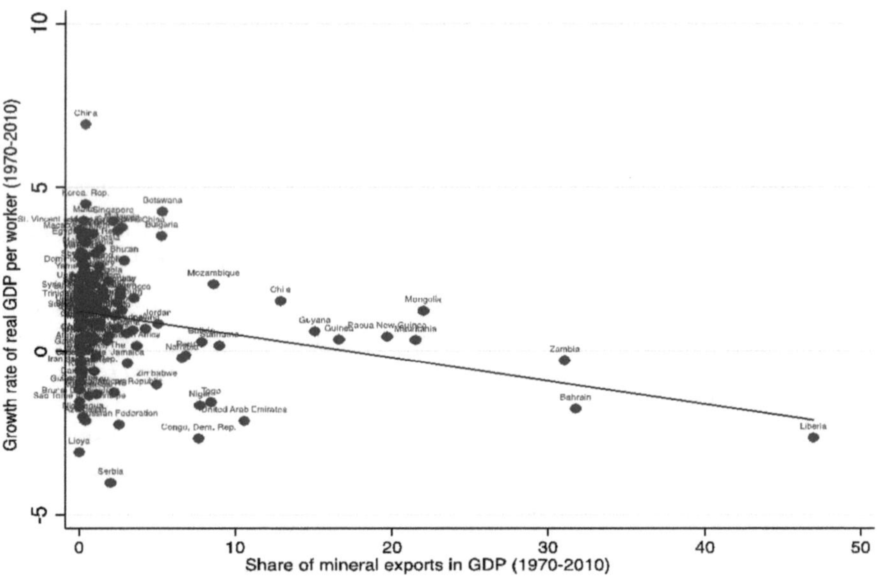

Fig. 3.2 Mineral exports and economic growth

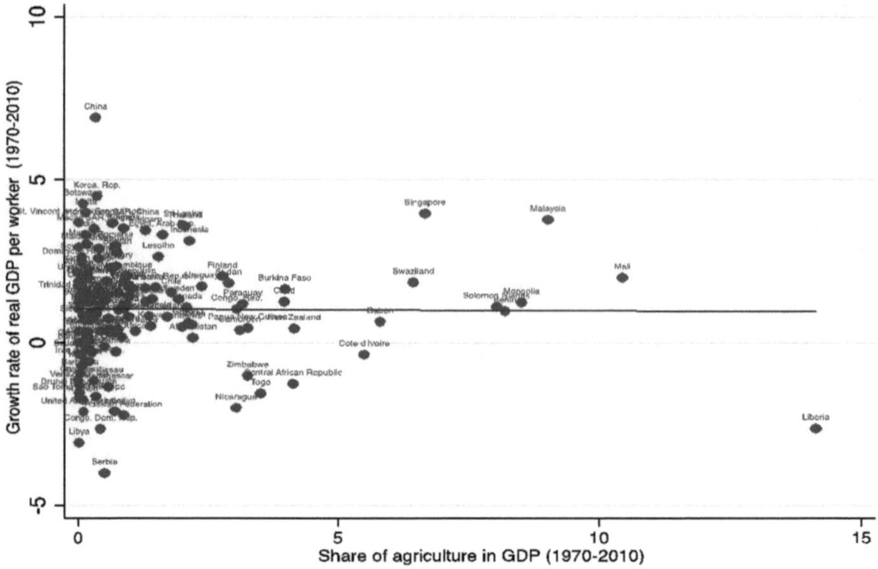

Fig. 3.3 Agricultural exports and economic growth

- X_i is a vector of the economic channels of the effect of natural resources on economic growth. The resource curse literature suggests that:
  - To measure growth in capital stock, the investment share of GDP (*Inv*) is used.
  - To capture the effect of the price volatility of natural resources on economic growth, a terms of trade variable (*TOT*) is included in the regression equation: the external *TOT* is the ratio of an export price index to an import price index.

The dependent variable and the main variables, selected on the basis of political and social transmission channels of the resource curse, are measured as follows:

- The dependent variable is the average annual growth, g_I, of adjusted per-person-employed GDP in 2005 international dollars of country i over the period 1970–2010. The analysis in this study uses GDP per worker, following Sachs and Warner (1995, 1997). It is argued that GDP per worker is more appropriate for growth models than GDP per capita because it encapsulates the productive population as a significant element of the production process, which is a key determinant in the growth model (Jones 1997; Durlauf et al. 2005).
- The political and social transmission channels of the resource curse are tested with institutional and human capacity measures. Institutional quality is measured by government effectiveness (*GEF*), accountability (*VA*), regulatory quality (*RQ*),

rule of law (*RL*), political stability (*PS*) and control of corruption (*CC*) rescaled to between 2.5 and −2.5. These are commonly used indicators of institutional quality in the resource curse literature. The variables are provided by the Work Bank (see the Appendix 1 for details). The percentage of the total population attaining secondary school education is a proxy for human capital (*HR*).

3.3.2 Measure of Resource Dependence

The second major decision relates to the measure of resource dependence. To measure the impact of natural resources on economic performance, it is important to distinguish between resource *abundance* and resource *dependence* (Brunnschweiler and Bulte 2008). Resource abundance is measured by total resource stock availability, such as total land area per capita, cropland per capita or mineral resources per capita (Auty 2001), while the term 'resource dependence' is used to describe the degree to which a natural resource-rich country is dependent on its primary product exports. Economic dependence on natural resources is measured by natural resource exports as a percentage of GDP or of total exports (Auty 2001; Ross 1999, 2012).

A resource-rich country's dependence on fuel (*FuelExp*), mineral (*MinExp*), and agricultural exports (*AgriExp*) is measured in terms of their respective shares of GDP. The ratio of primary exports to GDP used by Sachs and Warner (1997) and Ross (1999, 2012) is the most popular indicator for measuring an economy's resource dependence. In fact, this indicator is important because it shows, on the one hand, the role of natural resources in the export sector and, on the other hand, the significance of exports to the domestic economy (Ross 1999, 2012).

3.3.3 Different Types of Natural Resource

To shed more light on the effect on economic growth of different types of natural resource as primary commodities, this research investigates both 'point' resources, such as fuel and minerals, and 'diffuse' resources, such as agriculture (Auty 2001). Point resources are more problematic because they are usually capital intensive and controlled centrally by government (Auty 2001). Furthermore, the huge rents generated by point resources are easily appropriable as a source of rent-seeking and conflict, rather than a benefit to resource-abundant countries (Ross 2012; Karl 1999).

3.3.4 Cross-Sectional Model Versus Panel Data Model

In the first step, this study examines the effect of natural resources on economic growth, as well as its relationship with institutional quality and human resources, by constructing the interaction terms in the context of a cross-section of countries. The cross-sectional model has been employed by most empirical studies (e.g. Sachs and Warner 1995, 1997; Sala-i-Martin and Sulbramanian 2003; Mehlum et al. 2006; Boschini et al. 2007).

However, the cross-sectional method has some well-known drawbacks. On the one hand, this methodology is biased because of the unobserved heterogeneity of so-called omitted variables. On the other hand, being cross-sectional, the ordinary least squares (OLS) estimator relies entirely on a between-country comparison. This is misleading because of specific country characteristics not captured by other variables, such as culture, geographical features and political structure. Another significant drawback of the cross-sectional method is that cross-country regressions do not capture the effects of the dynamic behaviour of the data, for example the change over time of primary commodity prices such as crude oil, which have fluctuated greatly since 1970 and have a crucial effect on the economic growth of all oil-based economies.

Panel data offer a potential solution to the above-mentioned problems. This method makes more observations over longer time periods, which may diminish problems arising from omitted variable bias (Bond 2002). This research employs both methods: cross-sectional analysis using average growth rate for the 1970–2010 period; and a panel data analysis with data averaged in five-year periods to help to eliminate the negative influence of cyclical business fluctuations on the economic analysis (Durlauf et al. 2005).

3.4 Cross-Sectional OLS Regression

The cross-sectional regression model is based on Eq. 3.2.1 as follows[2]:

$$g_i = \beta_0 + \beta_1 g_{1970} + \beta_2 NR_i + \beta_3 Inv_i + \beta_4 ToT_i + \beta_5 HR_i + \beta_6 Inst_i + \varepsilon_i \qquad (3.4.1)$$

where the dependent variable is the average growth rate of GDP per worker g of country i over a forty-year period (between $t - 1$ and t) and g_{1970} is the log of per worker GDP in 1970 as the base year for the growth analysis as this was the first year of consistently available data for all countries. The ratio of natural resources exports to GDP (*NR*), including fuel, mineral and agricultural exports, investment as a share of GDP (*Inv*), the percentage of secondary school education attained in the total population (*HR*), and institutional quality (*Inst*), including government effectiveness, accountability, regulatory quality, rule of law, political stability, control of corruption and annual growth of terms of trade, are calculated over a forty-year period (1970–2010).

[2]The Stata software includes options with most rotines for estimating robust standard errors using the 'vce (rebust)' command.

The second stage estimates the interaction between natural resources and institutional quality and human capital using Eq. 3.4.2 to test whether the growth effects of natural resources are conditional on the pre-existing level of institutional quality and human resources in natural resource-based economies.

$$g_i = \beta_0 + \beta_i g_{it-1} + \beta_2 NR_i + \beta_3 Inv_i + \beta_4 ToT_i + \beta_5 HR_i + \beta_6 Inst_i + \beta_7 (NR * HR)_i + \beta_8 (NR * Inst)_i + \varepsilon_i \tag{3.4.2}$$

The effects of different types of natural resource (fuel, mineral and agricultural) exports on economic growth are reported as follows.

Table 3.5 provides results estimated by a simple OLS regression for a cross-section of countries over the period 1970–2010. In Columns 1 and 2, the average growth in GDP per worker is regressed on the share of fuel exports (*FEG*). In the following columns, the log of initial GDP per worker in 1970, investment, terms of trade, human resources and institutional quality (6-index) explanatory variables are added.

It can be seen that the coefficient of the initial GDP is always negative and highly significant, confirming the convergence predictions of neo-classical growth theory. The expected positive and significant impact of investment and education on economic growth is confirmed in Columns 1 and 2, which is consistent with the arguments of the Solow model that shows conditional convergence and a positive effect of education and investment on economic growth.

However, the terms of trade variable has a negative and statistically insignificant estimated coefficient. The high volatility of commodity prices causes terms of trade shocks, particularly in economies that specialise in commodities with less stable prices. This trade specialisation has a negative effect on growth (Van der Ploeg and Poelhekke 2009).

The government effectiveness variable appears statistically significant and its coefficient is positive, whereas the control of corruption variable is significant but its coefficient is negative. These results indicate that institutional quality is a significant growth determinant, as presented in the resource curse literature (Isham et al. 2005; Bulte et al. 2005; Ross 2012).

Turning to the coefficient of the fuel export variable, it is found that, after controlling for economic, social and institutional variables, it becomes positive and insignificant. To test the conditional hypothesis, interaction terms between the fuel exports variable and the statistically significant institutional indices (government effectiveness and control of corruption) are added, together with the human resources variable.

The conditional hypothesis suggested by Mehlum et al. (2006) and Boschini et al. (2007) states that quality of institutions plays a fundamental role in the effect of natural resources on economic performance. In estimating the conditional hypothesis interaction terms using Eq. (3.4.2), the interaction terms assert that the effect of a change in fuel exports on GDP growth in fuel-dependent countries is conditional on the quality of governmental effectiveness, control of corruption and

Table 3.5 Estimated effect of share of natural resource exports in GDP

Dependent variable: growth rate of GDP per worker (1970–2010)						
	Fuel exports		Mineral exports		Agricultural exports	
	(1)	(2)	(3)	(4)	(5)	(6)
Log GDP per worker 70	−1.2713***	−1.2679***	−1.1535***	−1.1875***	−1.1766***	−1.2120***
	(−9.23)	(−9.19)	(−9.54)	(−9.69)	(−9.18)	(−9.15)
Natural resource exports (NrExp)	0.0185	0.0870***	−0.0544***	0.0194	−0.0471	0.1539*
	(1.52)	(3.97)	(−4.63)	(0.55)	(−1.07)	(1.70)
Investment (Inv)	0.0372***	0.0259*	0.0392***	0.0426***	0.0358***	0.0389***
	(2.89)	(1.97)	(3.20)	(3.50)	(2.79)	(3.00)
Terms of trade (TOT)	−0.0107	−0.0316	0.0322	0.0343	0.0134	0.0169
	(−0.30)	(−0.87)	(1.07)	(1.13)	(0.40)	(0.52)
Human resource (HR)	0.0238***	0.0277***	0.0220***	0.0273***	0.0231***	0.0291***
	(3.75)	(3.93)	(3.53)	(3.84)	(3.64)	(3.85)
Voice and accountability (EstVA)	−0.1237	−0.2306	−0.2641	−0.2571	−0.2765	−0.2731
	(−0.50)	(−0.95)	(−1.35)	(−1.27)	(−1.29)	(−1.25)
Political stability (EstPS)	0.2541	0.3652**	0.2940*	0.2910*	0.2975	0.3039
	(1.36)	(2.05)	(1.68)	(1.67)	(1.62)	(1.63)
Government effectiveness (EstGEF)	1.5959***	1.6890***	1.3887***	1.3501***	1.4607***	1.2889**
	(3.13)	(3.03)	(2.90)	(2.78)	(2.84)	(2.13)
Regulatory quality (EstRQ)	0.0226	0.2384	0.1489	0.1790	0.0872	0.1109
	(0.06)	(0.67)	(0.46)	(0.54)	(0.26)	(0.33)
Rule of Law (EstRL)	0.1687	0.1613	0.2081	0.2082	0.3011	0.2542
	(0.32)	(0.30)	(0.42)	(0.42)	(0.58)	(0.48)
Control of corruption (EstCC)	−0.9752**	−1.2347***	−0.9302**	−1.0262**	−1.0210***	−0.9534*
	(−2.43)	(−2.93)	(−2.47)	(−2.54)	(−2.64)	(−1.90)
NrExp * EstGEF		−0.0612***		−0.0016		0.0801
		(−2.82)		(−0.04)		(0.83)
NrExp * EstCC		0.0586***		0.0388		0.0196
		(2.66)		(0.76)		(0.16)
NrExp * HR		−0.0010***		−0.0014***		−0.0042**
		(−3.01)		(−2.70)		(−2.39)
Intercept	10.3472***	10.3649***	9.5453***	9.4759***	9.7107***	9.6369***
	(9.71)	(9.66)	(10.41)	(10.40)	(9.71)	(9.53)
Countries	150	150	150	150	150	150
R-sq	0.573	0.622	0.604	0.615	0.565	0.573

t statistics in parentheses
Significant at *10%, **5% and ***1% level

human resource capacity. The results given in Table 3.5 show that the coefficient of the fuel export variable appears positive and significant after adding interaction terms.

The interaction effect with respect to the control of corruption variable outweighs the impact of fuel on fuel-dependent economies, and hence fuel resources tend to be positive for growth when corruption is controlled sufficiently. The estimate of interaction with government effectiveness and human resource variables has negative signs and is significant at the 1% level. This implies that dependence on natural resources contributes to economic growth only if institutional quality is high. This combined effect of a one-standard-deviation increase in the level of dependence on fuel exports and the quality of institutions (using government effectiveness) leads to a decline of 0.061 of a standard deviation in economic growth. In the same way, the combined effect of a one-unit increase in dependence on fuel exports and the quality of human resources leads to a 0.0001% drop in economic growth.

Table 3.5 also presents the estimated effect of mineral exports on economic growth. Column 3 shows that the coefficient of the initial GDP is always negative and highly significant. The effect of mineral exports on growth is negative and significant at the 1% level, but its effect does not remain significant in Column 4. The estimate of investment is positive and statistically significant and its effect is the same in Column 4. The effects of human resources and government effectiveness on economic growth are positive and highly statistically significant. However, the effect of corruption on economic growth is negative and statistically significant at the 1% level.

The results indicate that the interaction term is insignificant between mineral exports and both government effectiveness and control of corruption variables. However, the interaction between mineral exports and human resources is negative and highly significant at the 1% level. This illustrates that the effect of a one-unit change in mineral exports on economic growth is negative and 0.002% when the quality of human resources increases.

The findings relating to agricultural exports in Table 3.5 (Columns 5 and 6) show that the coefficient of the initial GDP is always negative and highly significant. However, agricultural exports have a negative and statistically insignificant relationship with economic growth. It can be seen that, in contrast to the corruption variable which is negative and significant at the 1% level, investment, human resources and government effectiveness are highly significant and positive. The estimate of interaction terms with government effectiveness and control of corruption is not significant, while with human resources it is negative and statistically significant at the 10% level. The coefficient is negative with respect to the interaction terms between agricultural exports and human resource capacity, indicating that a 1% increase in mineral exports brings about a decline in economic growth when human resource capacity increases.

3.4.1 Marginal Effect of Natural Resources on Economic Growth

To investigate the total effect of natural resource exports on economic growth, the following equation was used, composed of direct and indirect (interactive) effects of natural resource exports on growth:

$$\frac{\partial g_i}{\partial NR_i} = \beta_2 + \beta_7(HR_i) + \beta_8(Inst_i)$$

To calculate the total effect, which is estimated as a derivative of GDP growth with respect to fuel, mineral and agricultural exports, summary statistics (mean and standard deviation) are used, as shown in Table 3.1 and Columns 2, 4 and 6 of Table 3.5.

Marginal effects measure the impact of a one-unit change in one regressor in the dependent variables. If both regressors and dependent variable are in logs, the marginal effect measures the proportional change in percentage, which is equivalent to the elasticity. In the case of the growth regressions, the growth rate is measured as a change in logs while the institutional variable is measured by using a scale between − 1 and 1 (see Table 3.6). As a consequence, percentage changes or elasticities are not adequate, therefore it was used the effect of a unit change of the institutional variable (improvement in institutions) on economic growth (measured in percentage points).

Table 3.6 shows that a one-unit increase in fuel exports has no impact on economic growth as institutional and human resource quality improves. However, the effect of a one-unit increase in fuel exports is positive and 0.07% with low-quality institutions and human resources. As the average growth rate is one, a positive effect of 0.07 is approximately 1/14 of average growth. This represents a growth from 1 to 1.14 over a period of 40 years, which is inconsistent with the literature suggesting that resource-rich countries with poor institutions and human resource quality suffer from the curse (Mehlum et al. 2006).

Turning to the marginal effect of mineral exports on growth, each additional point of mineral exports leads to a 7% decline in economic growth for a high level of institutional and human resource quality. This means that a one-unit increase in mineral exports has a negative effect on economic growth as institutional and

Table 3.6 Marginal effects of share of natural resource exports on GDP per worker, 1970–2010

Marginal effects of natural resources in cross-sectional model		
Natural resource exports	High institutional quality = 1 High human resource quality = 87	Low institutional quality = −1 Low human resource quality = 25
Fuel exports	0	0.07
Mineral exports	−0.07	−0.05
Agricultural exports	−0.25	−0.12

human resource quality improves. The negative effect of −0.07 is around 1/14 of average growth, and shows a reduction in the growth rate from 1 to 0.86 over a period of 40 years. This outcome is consistent with the resource curse theory; however, it violates the conditional hypothesis.

The results also indicate that the estimate of the marginal effect of a one-unit increase in agricultural exports on economic growth associated with high levels of institutional and human resource quality is −0.25. This means that a 1% increase in agricultural exports leads to a decline in the growth rate from 1 to 0.96 over the period 1970–2010. These results do not confirm the conditional hypothesis.

A potential explanation that might be compatible with the main findings is that natural resource-dependent economies tend to reduce their economic dependence on the natural resource sector by diversifying their economic structure when institutional and human resource quality improve. This does not mean that economic growth will fall as a whole, because greater revenues will be generated from other sectors of the economy. The above results are supported by Meller's (2009) finding that natural resource-rich countries in Latin America became less dependent on natural resources over the period 1980–1990 by increasing the contribution of the manufacturing sector to economic growth.

The next section reports on testing to establish whether there is a difference between high and low natural resource-dependent economies in the effect of natural resources on growth. In order to do this, a dummy variable was used to split the sample into low and high natural resource-dependent countries. This was investigated only for fuel exports because dependent agricultural and mineral exports are small.

3.4.2 High Fuel-Dependent Economies

As stated earlier, the resource curse affects natural resource-rich countries that are heavily dependent on natural resource revenues (Karl 2004). Therefore, to examine more closely the effects of resource dependence on economic growth, the mineral-dependent index was used to identify fuel-dependent economies. Mineral- and fuel-rich countries whose mineral and fuel exports account for 8–10% of GDP are considered to be resource-dependent (Auty 1993; Gelb 1988; Davis 1995), although Auty (1990) proposed that a more comprehensive mineral-dependence index, accounting for the contribution of minerals to GDP, would be more than 20%.

Given this explanation, two dummy variables, D10 and D20, were created to identify how fuel resources influence economic growth in fuel-dependent economies. These new variables describe countries whose fuel export share of GDP is not less than 10 and 20% respectively and are defined as resource-dependent economies, and are used to estimate the following models:

$$g_i = \beta_0 + \beta_1 g_{i970} + \beta_2 FuelExp_i + \beta_3 Inv_i + \beta_4 TOT_i + \beta_5 (FuelExp * D_{10})_i + \beta_6 HR_i + \beta_7 EstVA_{it} + \beta_8 EstPS_i + \beta_9 EstGEF_i + \beta_{10} EstRQ_i + \beta_{11} EstRL_i + \beta_{12} EstCC_{it} + \varepsilon_{it} \quad (3.4.3)$$

$$g_i = \beta_0 + \beta_1 g_{i970} + \beta_2 FuelExp_i + \beta_3 Inv_i + \beta_4 TOT_i + \beta_5 (FuelExp * D_{10})_i + \beta_6 HR_i + \beta_7 EstAV_i + \beta_8 EstPS_i + \beta_9 EstGEF_i + \beta_{10} EstRQ_i + \beta_{11} EstRL_{it} + \beta_{12} EstCC_i + \beta_{13}(D_{10} * EstGEF)_i + \beta_{14}(D_{10} * EstCC)_i + \beta_{15}(D_{10} * HR)_i + \varepsilon_i \quad (3.4.4)$$

$$g_i = \beta_0 + \beta_1 g_{i1970} + \beta_2 FuelExp_i + \beta_3 Inv_i + \beta_4 TOT_i + \beta_5 (FuelEXP * D_{20})_i + \beta_6 HR_i + \beta_7 EstVA_i + \beta_8 EstPS_i + \beta_9 EstGEF_i + \beta_{10} EstRQ_i + \beta_{11} EstRL_i + \beta_{12} EstCC_i + \varepsilon_i \quad (3.4.5)$$

$$g_i = \beta_0 + \beta_1 g_{i1970} + \beta_2 FuelExp_i + \beta_3 Inv_i + \beta_4 TOT_i + \beta_5 (FuelExp * D_{20})_i + \beta_6 HR_i + \beta_7 EstAV_i + \beta_8 EstPS_i + \beta_9 EstGEF_i + \beta_{10} EstRQ_i + \beta_{11} EstRL_i + \beta_{12} EstCC_i + \beta_{13}(D_{20} * EstGEF)_i + \beta_{14}(D_{20} * EstCC)_i + \beta_{15}(D_{20} * HR)_i + \varepsilon_i \quad (3.4.6)$$

The above models are tested for the appropriateness of the specification by applying the Ramsey RESET. In other words, the results showed that the regressions are correctly specified by not rejecting the null hypothesis. Furthermore, because the models are nested, an F-test was used to compare the statistical the models. The F-test uses sums of squared residuals or a Likelihood Ratio test using likelihood values to select the accepted model specification to be used in the final analysis of the results. The test statistic of F = 7.4 (greater than the critical value of F-distribution = 4.3) led to the rejection of the null hypothesis so the second model (3.4.4) is statistically better than the first model (3.4.3). The F-test for the models 3.4.5 and 3.4.6 indicated that the second model (3.4.6) is the accepted model.

Table 3.7 shows that the estimated effect of a high degree of resource dependence on economic growth is positive but insignificant. The coefficient of the lower fuel export dependence (using 10% of fuel exports to GDP) is negative and insignificant. The total effects of fuel exports associated with each threshold of dependence are as follows:

$$FuelExport > 10\% = \beta_1 + \beta_2 = 0.0290 - 0.0062 = 0.22$$
$$FuelExport < 10\% = 0.03$$
$$FuelExport > 20\% = \beta_1 + \beta_2 = 0.0049 + 0.0401 = 0.05$$
$$FuelExport < 20\% = 0.005$$

Table 3.7 Estimated effect of fuel exports on GDP for dependent economies

Dependent variable: growth rate of GDP per worker (1970–2010)				
	(1)	(2)	(3)	(4)
Log GDP per worker 70	−1.2667***	−1.2521***	−1.2716***	−1.2381***
	(−9.18)	(−9.21)	(−9.20)	(−8.86)
Fuel export/GDP (FEG)	0.0451	0.0290	0.0049	0.0047
	(0.97)	(0.57)	(0.14)	(0.13)
FuelExp * D10	−0.0262	−0.0062		
	(−0.60)	(−0.12)		
EstGEF * D10		−2.4855**		
		(−2.49)		
EstCC * D10		1.8237*		
		(−1.93)		
HR * D10		−0.0045		
		(−0.61)		
FuelExp * D20			0.0132	0.0401
			(0.40)	(1.00)
EstGEF * D20				−2.7660**
				(−2.04)
EstCC * D20				2.1138
				(1.65)
HR * D20				−0.0224*
				(−1.89)
Investment (Inv)	0.0375***	0.0353***	0.0373***	0.0344***
	(2.90)	(2.85)	(2.89)	(2.75)
Terms of trade (TOT)	−0.0123	−0.0170	−0.0089	−0.0200
	(−0.34)	(−0.48)	(−0.25)	(−0.55)
Human resource (HR)	0.0236***	0.0218***	0.0243***	0.0237***
	(3.73)	(2.92)	(3.62)	(3.47)
Voice and accountability (EstVA)	−0.1242	−0.1746	−0.1440	−0.2253
	(−0.50)	(−0.70)	(−0.58)	(−0.88)
Political stability (EstPS)	0.2476	0.3105*	0.2603	0.3463*
	(1.30)	(1.73)	(1.37)	(1.84)
Government effectiveness (EstGEF)	1.5351***	1.7906***	1.6593***	1.7441***
	(2.74)	(3.14)	(2.84)	(2.92)
Regulatory quality (EstRQ)	0.0207	0.1599	−0.0091	0.1361
	(0.06)	(0.42)	(−0.02)	(0.34)
Rule of law (EstRL)	0.2104	0.1392	0.1615	0.1304
	(0.38)	(0.25)	(0.30)	(0.24)
Control of corruption (EstCC)	−0.9531**	−1.1524***	−1.0023**	−1.1421***
	(−2.33)	(−2.71)	(−2.50)	(−2.81)

(continued)

Table 3.7 (continued)

Dependent variable: growth rate of GDP per worker (1970–2010)				
	(1)	(2)	(3)	(4)
Intercept	10.2871***	10.3215***	10.3420***	10.1434***
	(9.61)	(9.85)	(9.76)	(9.54)
Countries	150	150	150	150
R-sq	0.574	0.605	0.574	0.603

t statistics in parentheses
Significant at *10%, **5% and ***1% level

The results indicate that the total effect of fuel exports in both groups is positive. However, the total effect of fuel exports using 20% of GDP is lower than for economies that are less dependent on fuel exports. With regard to the effect of institutional quality, the effect of government effectiveness (*EstGEF*) on growth is negative and insignificant for countries whose shares of fuel exports to GDP exceed both 10 and 20%. However, control of corruption (*EstCC*) appears positive and significant only for countries that are less dependent. The effect of human resource development is negative and significant at the 10% level in the highly dependent economies. The findings show that the institutional effect on economic growth is lower for countries with high dependence. This confirms the theory that the effect of fuel exports is lower in countries with poor institutions (Isham et al. 2005; Sala-i-Martin and Subramanian 2003; Mehlum et al. 2006; Boschini et al. 2007).

However, the results should be treated with caution since endogeneity problems remain because of potentially unobserved factors across countries that do not change over time and may cause a bias in model. To address these endogeneity issues, panel data were used, as described in the next section.

3.5 The Dynamic Econometric Model

Van der Ploeg and Poelhekke (2009) observe that empirical resource curse studies have used a variety of different methodologies to investigate growth determinants in resource-rich countries. Most studies have employed cross-sectional growth regression that has failed to consider the effect of commodity prices—namely, the dynamics of the resource curse. Therefore, they have suffered from endogeneity problems due to the omission of relevant variables. To remedy these problems, a wide range of empirical studies associated with economic growth has used the panel data method (Quah 1993; Islam 1995).

Bond (2002) argues that panel data allow for increased estimation efficiency by controlling individual effects. However, estimates based on fixed or random effects are biased and inconsistent when estimating a dynamic panel model. The argument for the inappropriateness of individual effects methods (fixed or random) to estimate a dynamic panel data model is that the error term and the lagged dependent variables are negatively correlated in the differentiating transformed equation. For instance, the following economic growth equations indicate that lagged GDP growth is correlated with the lagged error term after differencing the equation:

$$g_{it} - g_{it-1} = \beta_1(g_{it-1} - g_{it-2}) + \beta_2(X_{it} - X_{it-1}) + (\varepsilon_{it} - \varepsilon_{it-1})$$
$$g_{it} = \beta_0 + \beta_1 g_{it-1} + \beta_2 X_{it} + \mu_i + \varepsilon_{it}$$

g_{it-1} is correlated with $\mathcal{E}_{it-1}$; therefore, it is necessary to use instrument variables. g_{it-2} or $g_{it-2} - g_{it-3}$ may be used as instruments: both are valid since they are correlated with $g_{it-1} - g_{it-2}$ and uncorrelated with $\mathcal{E}_{it}$–$\mathcal{E}_{it-1}$ (Roodman 2006).

Another relevant argument, with which Lederman and Maloney (2008) concur, is that independent variables are not strictly exogenous, and hence are possibly correlated with past and current realisations of error terms. Generally, the context of a linear regression model is composed of one left-hand variable that is dynamic (depending on its own past realisation) and right-hand explanatory variables that are endogenous or predetermined, using generalised method of moments (GMM) estimators.

The predetermined variables are not strictly exogenous and correlated with past and possibly current realisations of the error term. This means that current and past shocks exert some feedback on the current values of the variables:

$$E[X_{it}, \varepsilon_{is}] \neq 0 \quad for\ s < t\, but \quad E[X_{it}, \varepsilon_{is}] = 0 \quad for\ s \geq t$$

The endogenous variables are correlated with the current error term:

$$E[X_{it}, \varepsilon_{is}] \neq 0 \quad for\, s \leq t\ \ but \quad E[X_{it}, \varepsilon_{is}] = 0 \quad for\ s > t$$

The GMM method described by Bond (2002) is useful for eliminating the above-mentioned econometric problems. A method generally referred to as the Arellano-Bond GMM is used to estimate the following dynamic panel equation:

$$g_{it} = \beta_0 + \beta_1 g_{it-1} + \beta_2 X_{it} + \mu_i + \varepsilon_{it} \quad i = 1, \ldots, N\, and\, t = 2, \ldots, T \qquad (3.5.1)$$

where i and t represent country and time periods respectively, and g_{it} and GDP_{it-1} are the contemporaneous and lagged realisations of the dependent variable, respectively. X_{it} is a vector of both exogenous and endogenous variables, $\mathcal{E}_{it}$ is a random error term and μ_i is the unobserved country-specific time invariant effect.

The unobserved country-specific time invariant effect is not observed by OLS methods because it is included in the residual term. Therefore, the error term (v_{it}) is composed of $\mathcal{E}_{it}$ and μ_i. Observation creates specific errors and unobserved country-specific time invariant effects that generate the following model:

$$g_{it} = \beta_0 + \beta_1 g_{it-1} + \beta_2 X_{it} + V_{it}, \quad V_{it} = \mu_i + \varepsilon_i \tag{3.5.2}$$

According to the assumptions of the GMM method, Model 3.5.2 violates the OLS estimations of strict exogeneity since the lagged dependent and independent variables are correlated with the fixed unobserved country effects and the lagged error term.

$$E(V_{it}|X_{it}, g_{it-1}) \neq 0$$

An alternative GMM technique suggests two steps to tackle the problems of omitted variable bias and endogeneity in order to obtain efficient and consistent estimators. The first step in GMM estimation deals with the unobserved country-specific effect μ_i by first differencing the model:

$$\begin{aligned} g_{it} &= \beta_0 + \beta_1 g_{it-1} + \beta_2 X_{it} + \mu_i + \varepsilon_{it} \quad i = 1,\ldots,N \quad and \ t = 2,\ldots,T \\ g_{it} - g_{it-1} &= \beta_1 (g_{it-1} - g_{it-2}) + \beta_2 (X_{it} - X_{it-1}) + (\varepsilon_{it} - \varepsilon_{it-1}) \quad i = 1,\ldots,N \quad and \quad t = 2,\ldots,T \\ \Delta g_{it} &= \beta_1 \Delta g_{it-1} + \beta_2 \Delta X_{it} + \Delta \varepsilon_{it} \quad i = 1,\ldots,N \quad and \ t = 2,\ldots,T \end{aligned} \tag{3.5.3}$$

As Arellano and Bond (1991) argue, the transformed Eq. 3.5.3 still violates the assumption of strict exogeneity. Therefore, they propose a second step, which is concerned with building valid instruments using g_{t-2} as an instrument for the lagged levels of the dependent, predetermined and endogenous variables. As a result, as long as the errors are not serially correlated, the method assumes:

$$\begin{aligned} E[g_{it-2}, \Delta \varepsilon_{it}] &= 0 \quad t \geq 3,\ldots,T \\ E[X_{it-2}, \Delta \varepsilon_{it}] &= 0 \quad t \geq 3,\ldots,T \end{aligned}$$

However, in the case of highly persistent variables across time, lagged levels tend to be weak instruments for differential variables. To overcome this issue, a system GMM estimator developed by Blundell and Bond (1998) is used. System GMM uses additional moment conditions in which lagged differences in explanatory variables as instrument level for equations in level. The additional moment conditions operate under the assumption that variables are not correlated with individual effect μ_i:

$$\begin{aligned} E[\Delta X_{it-s}, \mu_i] &= 0 \quad s \geq 1 \ t = 0,\ldots,T \\ E[\Delta g_{it-s}, \mu_i] &= 0 \quad s \geq 1 \ t = 0,\ldots,T \end{aligned}$$

As Roodman (2006, 2009) explains, the difference and system GMM method results in a more efficient outcome than the work of Arellano and Bover (1995). The Blundell-Bond system estimates equations through a combination of moment restrictions for differences and levels. Thus, this study uses the Blundell-Bond system GMM estimator, employing the Stata command 'xtabond2' provided by Roodman (2006, 2009) for estimations. As a rule of thumb, the difference GMM method requires the number of instruments used for estimations to be less than the number of cross-sectional units in the dataset (Roodman 2006, 2009). To keep the instrument count below the number of units, the instrument matrix is collapsed, as suggested by Roodman (2006, 2009). Further, the difference GMM has one- and two-step variants. Although the two-step variant is asymptotically more efficient and automatically produces standard errors that are robust to heteroscedasticity, these standard errors are typically downward biased (Arellano and Bond 1991).

In order to test the consistency and efficiency of the GMM estimator, three specification tests were employed: the Sargan/Hansen, AR (1) and AR (2) tests. The Sargan/Hansen test for over-identifying restrictions examines the null hypothesis of the validity of the set of instruments. Failure to reject this null hypothesis is therefore possible in the presence of a high p-value for the Sargan statistic; otherwise, the model or the instruments should be reconsidered. The AR (1) or Arellano-Bond test for autocorrelation proceeds under the null hypothesis that there is no correlation in the first-differenced residuals. The AR (2) or Arellano-Bond test for second-order autocorrelation is important, because the lagged values of the variables are used as instruments in order to test the following condition in which the disturbances are autocorrected up to an order greater than one; in other words, there is autocorrelation in levels:

$$\Delta\varepsilon_{it} = \varepsilon_{it} - \varepsilon_{it-1} \; and \; \Delta\varepsilon_{it-1} = \varepsilon_{it-1} - \varepsilon_{it-2} \text{ both have } \varepsilon_{it-1}$$

Therefore, the assumption of the null hypothesis is that there is no serial correlation for the second-order form. Furthermore, the residuals in the levels equation are not serially correlated. Thus, rejecting this null hypothesis implies model misspecification. As a general estimation strategy, the first equation is estimated in order to identify the economic growth determinants.

$$\begin{aligned} g_{it} = \beta_0 + \beta_i g_{it-1} + \beta_2 NR_{it} + \beta_3 Inv_{it} + \beta_4 ToT_{it} \\ + \beta_5 HR_{it} + \beta_6 Inst_{it}(6\,indices) + \mu_i + \varepsilon_{it} \end{aligned} \tag{3.5.4}$$

Therefore, the first baseline equation contains lagged GDP growth and the natural resource variables—fuel, mineral and agricultural exports of GDP (*NR*)—the economic variables, investment (*Inv*), terms of trade growth (*TOT*) and political and social variables, and the institutional variables (*Inst*) and human resource (*HR*). The interaction term between the natural resource variables and both institutional quality and human resource variables are added in a second model. The analytical aim of using different types of natural resource and their interaction with institutional and human resource quality is to identify the transmission channels of the resource curse.

$$g_{it} = \beta_0 + \beta_1 g_{it-1} + \beta_2 NR_{it} + \beta_3 Inv_{it} + \beta_4 TOT_{it} + \beta_5 HR_{it} + \beta_6 EstVA_{it} + \beta_7 EstPS_{it} + \beta_8 EstGEF_{it} + \beta_9 EstRQ_{it} + \beta_{10} EstRL_{it} + \beta_{11} EstCC_{it} + \mu_i + \varepsilon_{it} \quad (3.5.5)$$

$$g_{it} = \beta_0 + \beta_1 g_{it-1} + \beta_2 NR_{it} + \beta_3 Inv_{it} + \beta_4 TOT_{it} + \beta_5 HR_{it} + \beta_6 EstVA_{it} + \beta_7 EstPS_{it} + \beta_8 EstGEF_{it} + \beta_9 EstRQ_{it} + \beta_{10} EstRL_{it} + \beta_{11} EstCC_{it} + \beta_{12} (NR * Inst)_{it} + \beta_{13} (NR * HR)_{it} + \mu_i + \varepsilon_{it} \quad (3.5.6)$$

Applying the Ramsey RESET tested the appropriateness of the specification of the above models. The results showed that the regressions are correctly specified by not rejecting the null hypothesis. Furthermore, because the models are nested, an F-test was used to compare the models statistically (3.5.5 and 3.5.6). The F-test uses sums of squared residuals or the Likelihood Ratio test using likelihood values to select the accepted model specification to be used in the final analysis of the results. The test statistic of F = 9.3 (greater than the critical value of F-distribution = 6.1) resulted in the rejection of the null hypothesis so the second model (3.5.6) is statistically better than the first model (3.5.5). Both models described above investigate the potential determinants of economic growth, as well as the hypothesis of conditional convergence. However, they do not allow the economic growth rate during the transition to steady state to be subject to short-run business cycle fluctuations driven by trade shocks, which account for 45–60% of observed fluctuations in GDP (Mendoza 1995). Therefore, the first lags of natural resource, terms of trade and investment variables are added to Model 3.5.6 and are treated as predetermined. However, the lag in investment is omitted because it is not significant. In the same way, the lag in GDP growth is considered to be predetermined. Furthermore, predetermined variables X_i and $X_{it} - X_{it-1}$ will be correlated with $\varepsilon_{it} - \varepsilon_{it-1}$. In this case, $X_{i1} \ldots X_{it}$ are valid instruments, since they are uncorrelated with $\mathcal{E}_{it} - \mathcal{E}_{it-1}$ and correlated with $X_{it} - X_{it-1}$. Therefore, the momentary conditions for predetermined variables are:

$$E[GDP_{it-s}, \varepsilon_{it} - \varepsilon_{it-1}] = 0 \quad t \geq 2 \quad and \quad 1 \leq s < t$$
$$E[NR_{it-s}, \varepsilon_{it} - \varepsilon_{it-1}] = 0 \quad t \geq 2 \quad and \quad 1 \leq s < t$$
$$E[TOT_{it-s}, \varepsilon_{it} - \varepsilon_{it-1}] = 0 \quad t \geq 2 \quad and \quad 1 \leq s < t$$

$$g_{it} = \beta_0 + \beta_1 g_{it-1} + \beta_2 NR_{it} + \beta_3 NR_{it-1} + \beta_4 Inv_{it} + \beta_5 TOT_{it} + \beta_6 TOT_{it-1} + \beta_7 HR_{it} + \beta_8 Inst_{it} + \beta_9 (NR * HR)_{it} + \beta_{10} (NR * Inst)_{it} + \beta_{11} (NR_{it-1} * HR_{it}) + \beta_{12} (NR_{it-1} + Inst_{it}) + \mu_i + \varepsilon_{it} \quad (3.5.7)$$

In using the model for system GMM, the different natural resource variables, growth in terms of trade, investment, institutional indices and human resources are treated as endogenous. Moreover, independent variables X_i are endogenous: X_{it}–X_{it-1} is correlated with $\mathcal{E}_{it} - \mathcal{E}_{it-1}$ because X_{it} is correlated with both $\mathcal{E}_{it}$ and $\mathcal{E}_{it-1}$. In this case, $X_{i1} \ldots X_{it-2}$ are valid instruments since they are uncorrelated with $\mathcal{E}_{it} - \mathcal{E}_{it-1}$ and correlated with $X_{it} - X_{it-1}$. Therefore, the momentary conditions for endogenous variables are as follows:

$$\begin{array}{llll} E[NR_{it-s}, \varepsilon_{it} - \varepsilon_{it-1}] = 0 & t \geq 3 & and & 2 \leq s < t \\ E[TOT_{it-s}, \varepsilon_{it} - \varepsilon_{it-1}] = 0 & t \geq 3 & and & 2 \leq s < t \\ E[Inv_{it-s}, \varepsilon_{it} - \varepsilon_{it-1}] = 0 & t \geq 3 & and & 2 \leq s < t \\ E[Inst_{it-s}, \varepsilon_{it} - \varepsilon_{it-1}] = 0 & t \geq 3 & and & 2 \leq s < t \\ E[HR_{it-s}, \varepsilon_{it} - \varepsilon_{it-1}] = 0 & t \geq 3 & and & 2 \leq s < t \end{array}$$

The baseline mode tested for the correct functional form for variables by applying the Ramsey RESET. The test showed that there are no omitted-nonlinear-variables to include in the models. The model is run with the system GMM, using STATA, comprising the two-step difference GMM, Windmeijer-correlated standard errors, small-sample adjustments and orthogonal deviations (Roodman 2009).

$$\begin{array}{l} Xtaband2: GDPgrowth\, L1.GDPgrowth\, NR\, L1.NR\, Inv\ TOT\ L1.TOT\ HR\ EstAV \\ EstPS\ EstGEF\ EstRL\ EstCC\ NR * Inst\quad i,t,\ gmmstyle\,(L1.GDPgrowth\, L.NR\, L1.NR\, L.TOT \\ L1.TOT\, L.Inv\, L.HR\, L.EstAV\, L.Estps\, L.EstGEF\, L.EstRQ\, L.EstRL\, L.EstCC\, L.(NR * Inst)) \\ ivstyle(t)\ two - step\ robust\ small\ orthogonal \end{array}$$

3.5.1 Estimation Results

The estimated results for the share of resource exports in GDP for different types of aggregate resources are as follows. Fuel, mineral, and agricultural exports are reported in Tables 3.8, 3.9 and 3.10 for the dynamic panel model specification. The output of the system GMM associated with all resource exports is consistent and efficient due to rejection of the null hypothesis of over-identifying restrictions as valid.

Table 3.8 indicates that the lag sign for GDP is insignificant, which is inconsistent with the convergence predictions of neo-classical growth theory. The contemporaneous coefficient estimate of fuel exports is out of line with the existing empirical resource curse literature. It also appears to be insignificant over the next five years. The investment variable is positive and significant at the 1% level, which shows a high impact on growth rate of a 1% increase in the investment rate. The sign of the coefficient associated with the terms of trade variable in the current period appears significant at the 10% level. However, the effect of terms of trade on economic growth five years later is insignificant. Of the six dimensions of

Table 3.8 Estimated effect of share of fuel exports in GDP per worker (system GMM panel data estimates)

Dependent variable: growth rate of GDP per worker (1970–2010)			
	(1)	(2)	(3)
GDP growth t − 1	−0.0083	−0.0287	0.0224
	(−0.16)	(−0.40)	(0.34)
Fuel export/GDP (FEG)	0.0169	−0.0357	0.0131
	(0.21)	(−0.69)	(0.19)
FuelExp * EstRQ		0.0915	0.1083
		(1.26)	(0.66)
FuelExp * EstPS		−0.0058	0.0446
		(−0.15)	(0.64)
FuelExp * EstRL		−0.0942	−0.1635
		(−1.10)	(−0.85)
Fuel Export/GDP t − 1 (FEG)	−0.0770	−0.0475	−0.0746
	(−1.04)	(−0.98)	(−1.16)
FuelExp t – 1 * EstRQ			−0.0931
			(−0.64)
FuelExp t – 1 *EstPS			−0.0499
			(−0.70)
FuelExp t – 1 * EstRL			0.1485
			(0.86)
Investment (Inv)	0.1141***	0.0892**	0.0773**
	(3.07)	(2.00)	(1.99)
Terms of trade (TOT)	0.0208*	0.0183*	0.0135
	(1.80)	(1.66)	(1.04)
Terms of trade t − 1 (TOT)	0.0037	0.0060	0.0076
	(0.21)	(0.50)	(0.43)
Human resource (HR)	−0.0016	0.0205	0.0135
	(−0.10)	(1.16)	(0.80)
Voice and accountability (EstVA)	−0.1875	−0.5496	−0.4407
	(−0.21)	(−0.74)	(−0.60)
Political stability (EstPS)	0.9103*	0.7567	0.8438
	(1.77)	(1.25)	(1.58)
Government effectiveness (EstGEF)	−0.9014	−0.7425	−0.5301
	(−0.77)	(−0.70)	(−0.50)
Regulatory quality (EstRQ)	2.7987***	1.8564**	2.0118**
	(2.65)	(2.04)	(2.34)
Rule of law (EstRL)	−2.7261*	−1.8391	−1.9705
	(−1.86)	(−1.42)	(−1.44)
Control of corruption (EstCC)	0.4697	0.3223	−0.1247
	(0.46)	(0.32)	(−0.13)

(continued)

Table 3.8 (continued)

Dependent variable: growth rate of GDP per worker (1970–2010)			
	(1)	(2)	(3)
Intercept	−3.7788***	−4.6271***	−4.2836***
	(−2.62)	(−3.09)	(−2.83)
Observations	324	324	324
Countries	132	132	132
AR (2) (*P*-value)	0.317	0.085	0.089
Hansen test (*P*-value)	0.986	0.871	0.966

t statistics in parentheses
Significant at *10%, **5% and ***1% level

institutional quality, three—political stability, regulatory quality and rule of law—have significant effects on economic growth. The interaction of these significant measures of institutional quality with fuel exports is added into the model for estimation. The effect of interaction terms on economic growth is not significant. In the next step, the effect of the degree of fuel dependence on economic growth in fuel-dependent economies was tested, based on the two fuel-dependence indices.

The results from regression of GDP per worker on mineral exports show that the effect of mineral exports is not statistically significant. As can be seen in Table 3.9, the estimate of lagged GDP is insignificant. The estimate of investment and terms of trade is positive and significant at the 10% level. A 1% increase in investment increases economic growth by 0.09%, and a 1% increase in terms of trade increases growth by 0.02%. The effects of both voice and accountability and regulatory quality variables on growth are significant and positive. The coefficient of the rule of law variable is negative and economically significant at the 5% level. To test the conditional hypothesis, the interaction of these significant measures of institutional quality with mineral exports is added into the model and is not significant.

Table 3.10 indicates that the current effect of agricultural exports on growth is negative and statistically significant at the 5% level. This is consistent with the resource curse hypothesis. However, the coefficient of lagged agricultural exports is positive and also significant at the 1% level, suggesting that a one-unit increase in future agricultural exports leads to a 0.037% rise in economic performance, and to 0.09% in the long run. The effect of investment on growth is positive and significant at the 0.01% level. The coefficient of both voice and accountability and terms of trade variables have a positive sign and are statistically significant at the 10% level. After controlling the interaction term, agricultural exports lose their significant effect on economic growth in the current and next five-year period. The estimate of the interaction term between institutional quality (voice and accountability variable) and agricultural exports appears insignificant, which does not confirm the conditional hypothesis.

Table 3.9 Estimated effect of share of mineral exports on GDP per worker (system GMM panel data estimates)

Dependent variable: growth rate of GDP per worker (1970–2010)			
	(1)	(2)	(3)
GDP growth t − 1	0.0244	−0.0520	0.0009
	(0.34)	(−0.79)	(0.01)
Mineral Exp/GDP (MEG)	−0.0853	−0.2663	−0.1103
	(−0.84)	(−1.16)	(−1.09)
MinExp * EstAV		0.3747*	0.1156
		(1.85)	(0.60)
MinExp * EstRQ		−0.1352	−0.1083
		(−0.38)	(−0.34)
MinExp * EstRL		−0.1894	0.0170
		(−0.45)	(0.05)
Mineral Exp/GDP t − 1 (MEG)	−0.0258	0.1137*	0.0475
	(−0.35)	(1.89)	(0.51)
MinExp t − 1 * EstAV			−0.1013
			(−1.26)
MinExp t − 1 * EstRQ			−0.2068
			(−0.64)
MinExp t − 1 * EstRL			0.1310
			(0.39)
Investment (Inv)	0.0902*	0.0915*	0.0776
	(1.89)	(1.82)	(1.31)
Terms of trade (TOT)	0.0222**	0.0327***	0.0287***
	(2.40)	(3.50)	(2.72)
Human resource (HR)	−0.0117	0.0102	0.0166
	(−0.70)	(0.33)	(0.60)
Voice and accountability (EstVA)	1.0777*	0.5849	1.2534
	(1.69)	(0.66)	(1.29)
Political stability (EstPS)	0.6044	1.0467	1.5999*
	(1.01)	(1.19)	(1.91)
Government effectiveness (EstGEF)	−0.8613	0.2186	−0.1166
	(−0.68)	(0.11)	(−0.06)
Regulatory quality (EstRQ)	2.6190**	3.1981	3.8922*
	(2.09)	(1.47)	(1.87)
Rule of law (EstRL)	−2.7705*	−3.3089	−5.0452*
	(−1.66)	(−1.47)	(−1.95)
Control of corruption (EstCC)	0.1692	−0.7803	−0.0808
	(0.11)	(−0.46)	(−0.05)
Intercept	−3.5019**	−5.8387***	−5.8908***
	(−2.59)	(−2.82)	(−2.97)

(continued)

Table 3.9 (continued)

Dependent variable: growth rate of GDP per worker (1970–2010)			
	(1)	(2)	(3)
Observations	389	389	389
Countries	137	137	137
AR (2) (*P*-value)	0.153	0.2	0.152
Hansen test (*P*-value)	0.243	0.425	0.354

t statistics in parentheses
Significant at *10%,**5% and ***1% level

3.5.2 Long-Term Effect of Natural Resource Exports on Growth

To investigate the long-run effect of natural resource exports on economic growth, the following equation was used, composed of direct and indirect (interactive) effects on growth in natural resource exports in the current and previous five-year periods:

$$\frac{\partial g_{it}}{\partial NR_{it}} = \beta_2 + \beta_9(HR_{it}) + \beta_{10}(Inst_{it})$$

$$\frac{\partial g_{it}}{\partial NR_{it_{-1}}} = \beta_3 + \beta_{11}(HR_{it}) + \beta_{12}(Inst_{it})$$

To calculate the total effect (Table 3.11), estimated as a derivative of GDP growth with respect to fuel, mineral and agricultural exports, the summary statistics (mean and standard deviation) in Table 3.1 and the last regressions of Tables 3.8, 3.9 and 3.10 were used.

The outcome indicates that estimates of the long-range effect of a one-unit increase in fuel exports on economic growth with good institutional quality are negative. Estimates of the long-term effect of a one-unit increase in mineral exports on economic growth for a high level of institutional quality are also negative. Neither set of results confirms the conditional hypothesis, but both are consistent with the resource curse theory.

In Table 3.11, the outcome indicates that estimates of the long-term effect of a one-unit increase in agricultural exports on economic growth with a high level of institutional quality are positive. This implies that the diverse effects of the resource curse may be mitigated through the improvement of institutional quality, which is consistent with the conditional hypothesis. Furthermore, The short and long marginal effects changes across countries when one adds interaction terms to the institutional variables. As a consequence, short and long run effects changes across countries depending on their institutional variable levels. Table 3.11 indicates that the short and long-run marginal effects vary for countries with low and high fuel-dependence as countries have different values of institutional quality.

Table 3.10 Estimated effect of share of agricultural exports on GDP per worker (system GMM panel data estimates)

Dependent variable: growth rate of GDP per worker (1970–2010)		
	(1)	(2)
GDP growth t − 1	0.0295	0.0427
	(0.44)	(0.64)
Agricultural export/GDP (AEG)	−0.4563**	−0.2306
	(−2.56)	(−1.23)
Agricultural export/GDP t − 1 (AEG)	0.3665***	0.2102
	(2.79)	(1.26)
Investment (Inv)	0.1281***	0.1110**
	(2.62)	(2.52)
Terms of trade (TOT)	0.0162*	0.0117
	(1.84)	(1.24)
Human resource (HR)	−0.0144	−0.0080
	(−0.95)	(−0.44)
Voice and accountability (EstVA)	1.1585*	0.6572
	(1.72)	(0.82)
Political stability (EstPS)	0.4520	0.5076
	(0.79)	(0.89)
Government effectiveness (EstGEF)	−0.0577	−0.8989
	(−0.04)	(−0.72)
Regulatory quality (EstRQ)	1.6268	2.0348*
	(1.30)	(1.90)
Rule of law (EstRL)	−2.1568	−2.3708*
	(−1.50)	(−1.95)
Control of corruption (EstCC)	−0.3617	0.4593
	(−0.32)	(0.45)
AEG * EstAV		0.5679
		(1.23)
AEG t – 1 * EstAV		−0.3173
		(−0.82)
Intercept	−3.2659**	−2.7684
	(−2.53)	(−1.65)
Observations	388	388
Countries	137	137
AR (2) (*P*-value)	0.304	0.251
Hansen test (*P*-value)	0.441	0.376

t statistics in parentheses
Significant at *10%, **5% and ***1% level

Table 3.11 Long-run effects of natural resources with institutional quality

Natural resource exports	Current effect	Effect five years later	Long-run effect
Long-run effects of natural resources with high institutional quality = 1			
Fuel exports	0.003	0.07	−0.07
Mineral exports	0.04	−0.13	−0.10
Agricultural exports	0.34	−0.11	0.24
Long-run effects of natural resources with low institutional quality = −1			
Fuel exports	0.023	−0.08	−0.06
Mineral exports	−0.23	0.23	0
Agricultural exports	−0.80	0.53	−0.27

Table 3.12 Distribution and confidence interval of different groups

Variables	Mean	Standard error	[95% conf. interval]
FuelExp * D20	3.94	0.41	3.14–4.75
FuelExp * D10	4.78	0.42	4–5.6

3.5.3 Fuel-Dependent Countries

As explained in the cross-sectional model,[3] dummy variables *D10* and *D20* were used to investigate the economic performance of fuel economies with fuel exports more than 10% (*D10*) and 20% (*D20*) of GDP, see Table 3.13.

As explained in Sect. 3.4.2, different thresholds for fuel export dependency are used that are based on the resource curse literature. However, Tables 3.12 and 3.13 show that the differences between the effects of fuel export for two groups (*D*10 and *D*20) on economic growth is negligible. Furthermore, the results presented in Table 3.13 show that in fuel-dependent countries (using *D10*), the effect of fuel exports on economic performance is positive (0.4303 − 0.2591 = 0.17) in the current period, but appears negative (−0.5599 + 0.2910 = −0.27) five years later. The long-term effect of fuel exports in dependent economies, taking into account dummy variable *D10*, is (0.17 − 0.27)/(1 − 0.0196) = −0.1. A 1% increase in fuel exports (using *D10*) leads to a 0.1% decline in economic growth. This demonstrates the existence of the resource curse in countries where fuel exports form more than 10% of GDP.

It is evident from Column 2 of Table 3.13 that the effect of institutional quality (regulatory quality) is positive and significant at the 1% level for economies that are less dependent on fuel exports. However, institutional quality has less effect on economic growth in dependent economies. The sign of the coefficient relating to the

[3]Another common way of capturing non-linearity in the relationship, in addition to the interactive effects, is to add squares of the key variables. The negative/positive sign of interaction will indicate substitution/complementarity among the input variables. The squares indicate non-linearity in the relationship. The non-linearity can be direct or indirect.

Table 3.13 Estimated effect of fuel exports in low- and high-dependence countries (system GMM panel data estimations)

Dependent variable: growth rate of GDP per worker (1970–2010)				
	(1)	(2)	(3)	(4)
GDP growth t − 1	0.0053	−0.0196	0.0087	−0.0063
	(0.07)	(−0.22)	(0.12)	(−0.08)
Fuel export/GDP (FEG)	0.2979	0.4303*	0.1878	0.2936
	(1.61)	(1.72)	(1.12)	(1.59)
Fuel export/GDP t − 1 (FEG)	−0.5247***	−0.5599**	−0.1801	−0.2225
	(−2.85)	(−2.10)	(−1.38)	(−1.26)
FuelExp * D10	−0.1444	−0.2591		
	(−0.84)	(−1.10)		
FuelExp * D10 t − 1	0.2896*	0.2910		
	(1.78)	(1.27)		
EstRQ * D10		−1.7361		
		(−1.13)		
FuelExp * D20			−0.0933	−0.1829
			(−0.66)	(−1.44)
FuelExp * D20 t − 1			−0.0074	−0.0249
			(−0.10)	(−0.16)
EstPS * D20				−4.3052
				(−1.28)
EstRQ * D20				−1.5051
				(−0.36)
EstRL * D20				5.6440
				(1.49)
Investment (Inv)	0.1510***	0.1589***	0.1583**	0.1704***
	(3.17)	(2.89)	(2.45)	(2.90)
Terms of trade (TOT)	−0.0042	−0.0023	0.0058	0.0031
	(−0.27)	(−0.10)	(0.33)	(0.18)
Terms of trade t − 1 (TOT)	0.0311**	0.0351**	0.0231	0.0211
	(2.28)	(2.38)	(1.26)	(1.42)
Human resource (HR)	0.0206	0.0112	0.0034	0.0052
	(0.86)	(0.39)	(0.11)	(0.19)
Voice and accountability (EstVA)	−0.8046	−0.8674	−0.4758	−0.7081
	(−0.99)	(−0.88)	(−0.46)	(−0.67)
Political stability (EstPS)	1.0922	0.8183	1.4711*	1.5395
	(1.35)	(1.12)	(1.86)	(1.53)
Government effectiveness (EstGEF)	−1.4636	−0.7031	0.0084	−0.5505
	(−0.96)	(−0.48)	(0.00)	(−0.30)
Regulatory quality (EstRQ)	3.7429**	4.1992***	4.1530**	3.9050*
	(2.59)	(2.83)	(2.44)	(1.96)

(continued)

Table 3.13 (continued)

Dependent variable: growth rate of GDP per worker (1970–2010)				
	(1)	(2)	(3)	(4)
Rule of law (EstRL)	−3.2735*	−3.2489	−4.0549*	−3.3862
	(−1.69)	(−1.56)	(−1.82)	(−1.42)
Control of corruption (EstCC)	0.2493	−0.5133	−0.8745	−0.6820
	(0.17)	(−0.29)	(−0.50)	(−0.48)
Intercept	−6.7467***	−7.2339***	−6.6707***	−6.4693***
	(−3.73)	(−3.56)	(−3.19)	(−3.14)
Observations	324	324	324	324
Countries	132	132	132	132
AR (2) (*P*-value)	0.317	0.27	0.162	0.207
Hansen test (*P*-value)	0.986	0.881	0.954	0.953

t statistics in parentheses
Significant at *10%, **5% and ***1% level

human resource variable is positive but insignificant for economies both dependent and non-dependent on fuel exports. Column 4 indicates the insignificant effect of institutions and human resources on growth for countries highly dependent on fuel exports (using *D20*).

Column 4 also indicates that a one-unit increase in fuel exports in countries where fuel revenues are the main source of national income leads to a 0.11% increase in economic growth in the current period, with a 0.25% reduction in growth in the next five years. Its long-run effect on economic performance is also negative $(-0.25 + 0.11)/(1 - 0.0063) = -0.14$.

3.6 Conclusion

Using the cross-sectional model, the set of regressions shows that the effect of the different types of natural resource is positive on growth after controlling the interaction terms. However, the effect of the different dimensions of institutional quality on growth varies. For example, the impact of control of corruption (*EstCC*) is positive and significant with respect to fuel exports, whereas effectiveness of government (*EstGEF*) has a negative impact on growth, while other dimensions remain insignificant. Therefore, the estimations of interactions also vary.

The main findings indicate that both fuel and mineral exports delay economic growth, conditional on low institutional quality (using control of corruption). In other words, most natural resource-dependent economies, especially oil and gas exporters, suffer from the resource curse when their wealth is combined with poor-quality human and institutional capacity, resulting in a high level of dependence on natural resource exports (Brunnschweiler and Bulte 2008).

The marginal effects of all types of natural resource are negative, evaluated in terms of high and low levels of institutional and human resource quality. The results show that the effect of fuel exports on economic growth will be negligible with high-quality institutions and human resources. The marginal effect of mineral and agricultural exports on economic performance is negative when institutional and human resource quality improves. This is compatible with Meller's (2009) finding that resource-rich countries in Latin America tend to reduce their economic dependence on natural resources by diversifying their export structure, thus increasing non-resource exports. This leads to a reduction in the impact of natural resources on economic growth.

However, using the panel data model, the findings indicate that the long-term effect of fuel and mineral exports is negative, thus confirming the existence of the resource curse in fuel- and mineral-rich countries. Conversely, the long-run effect of agricultural exports on growth is positive. This means that countries abundant in point resources are more prone to the resource curse than diffuse resource-abundant countries. In contrast to the cross-sectional model, the interaction terms between all natural resources and economic growth are not significant using the panel data. In other words, the results of the dynamic panel model do not confirm the hypothesis that the effect of natural resources is conditional on the quality of institutions.

The results of both cross-sectional and panel data models indicate that high economic dependence on fuel exports leads to lower economic performance. The main finding associated with the effect of fuel exports on economic growth is that fuel exports have an adverse effect on growth only for countries that are heavily dependent on fuel exports. Moreover, the effect of institutions is lower in highly fuel-dependent countries than in those with economies less reliant on fuel export revenues. As a result, the economic performance of fuel-dependent economies is more likely to suffer from the detrimental effects of the resource curse (Karl 2004; Auty 1993; Ding and Field 2005).

These results should be treated with caution since limited data are available and the various dimensions of institutional quality are subject to measurement error. As Ross (2014) argues, definitions of institutional quality indices are ambiguous, and institutions may also be affected by natural resources. This makes it difficult to draw conclusions about the possible existence of a causal relationship between fuel exports and institutional quality. Moreover, how natural resources are measured may affect the results, including the distinction between natural resource dependence and natural resource abundance, and how they are gauged.

In this study, the empirical results show the importance of economic, political and social transmission mechanisms of the resource curse with regard to the different types of natural resource, particularly in fuel-exporting countries, which is the main focus here. Comparative studies of resource-rich countries will be necessary to shed light on causal explanations for poor economic growth through these transmission channels and highlight causal explanations for economic, political and social channels in which natural resource dependence may affect economic growth;

in other words, to help draw policy lessons from the experience of other resource-based industries to tackle the diverse curse effects of petroleum revenues on long-run economic development.

Appendix 1: Institutional Indices Definition

1. **Voice and Accountability (VA)**: Reflects perceptions of the extent to which a country's citizens are able to participate in selecting their government, as well as freedom of expression, freedom of association, and a free media.
2. **Political Stability (PS)**: Reflects perceptions of the likelihood that the government will be destabilized or overthrown by unconstitutional or violent means, including politically motivated violence and terrorism.
3. **Government Effectiveness (GEF)**: Reflects perceptions of the quality of public services, the quality of the civil service and the degree of its independence from political pressures, the quality of policy formulation and implementation, and the credibility of the government's commitment to such policies.
4. **Regulatory Quality (RQ)**: Reflects perceptions of the ability of the government to formulate and implement sound policies and regulations that permit and promote private sector development.
5. **Rule of Law (RL)**: Reflects perceptions of the extent to which agents have confidence in and abide by the rules of society, and in particular the quality of contract enforcement, property rights, the police, and the courts, as well as the likelihood of crime and violence.
6. **Control of Corruption (CC)**: Reflects perceptions of the extent to which public power is exercised for private gain, including both petty and grand forms of corruption, as well as "capture" of the state by elites and private interests.

References

Arellano M, Bond S (1991) Some tests of specification for panel data: monte carlo evidence and an application to employment equations. Rev Econ Stud 58(2):277–297

Arellano M, Bover O (1995) Another look at the instrumental variable estimation of error-components models. J Econom 68(1):29–51

Auty RM (1990) Resource-based industrialization: sowing the oil in eight developing countries. Oxford University Press, Oxford

Auty RM (1993) Sustaining development in mineral economies: the resource curse book. Routledge, London

Auty RM (ed) (2001) Resource abundance and economic development. Oxford University Press, Oxford

Blundell R, Bond S (1998) Initial conditions and moment restrictions in dynamic panel data models. J Econom 87:115–143

Bond SR (2002) Dynamic panel data models: a guide to micro data methods and practice. Port Econ J 1(2):141–162

Boschini AD, Petterson J, Roine J (2007) Resource curse or not: a question of appropriability. Scand J Econ 109(3):593–617
Brunnschweiler CN, Bulte EH (2008) The resource curse revisited and revised: a tale of paradoxes and red herrings. J Environ Eco Manag 55:248–264
Bulte EH, Damania R, Deacon RT (2005) Resource intensity, institutions, and development. World Dev 33(7):1029–1044
Davis GA (1995) Learning to love the Dutch disease: evidence from the mineral economies. World Dev 23(10):1765–1779
Ding N, Field BC (2005) Natural resource abundance and economic growths. Land Econ 81 (4):496–502
Durlauf SN, Johnson PA, Temple JR (2005) Growth econometrics. In: Aghion P, Durlauf SN (eds) Handbook of economic growth, vol 1, pp 555–677
Gelb, A (1988) Oil windfalls: blessing or curse? Oxford University Press,Oxford
Isham J, Woolcock M, Pritchett L, Busby G (2005) The varieties of resource experience: natural resource export structures and the political economy of economic growth. World Bank Econ Rev 19(2):141–174
Islam N (1995) Growth empirics: a panel data approach. Q J Econ 110(4):1127–1170
Jones CI (1997) On the evolution of the world income distribution. J Econ Perspect 11(3):19–36
Karl TL (1999) The perils of the petro-state: reflections on the paradox of plenty. J Int Aff 53(1): 31–38
Karl TL (2004) Oil-led development: social, political, and economic consequences. In: Cleveland CJ (ed) Encyclopaedia of energy. Elsevier, Amsterdam, pp 661–672
Lederman D, Maloney WF (2008) In search of the missing resource curse. Mimeograph., The World Bank, Washington, DC
Lee K, Pesaran MH, Smith RP (1997) Growth and convergence in a multi-country empirical stochastic Solow model. J Appl Econ 12(4):357–392
Mehlum H, Moene K, Torvik R (2006) Institutions and the resource curse. Econ J 116(508):1–20
Meller P (2009) From unilateral liberalization to regional free trade agreements: a Latin America perspective. Econ Change Restruc 42(1):85–103
Mendoza EG (1995) The terms of trade, the real exchange rate, and economic fluctuations. Int Econ Rev 36(1):101–137
Quah D (1993) Empirical cross-section dynamics in economic growth. Eur Econ Rev 37(2–3): 426–434
Roodman D (2006) How to do xtabound2: an introduction to 'difference' and 'system' GMM in Stata. Working paper 103. Center for global development, Washington, DC
Roodman D (2009) Practitioners' corner: a note on the theme of too many instruments. Oxford Bull Econ Stat, 71(1). doi:10.1111/j.1468-0084.2008.00542.x
Ross ML (1999) The political economy of the resource curse. World Polit 51(2):297–322
Ross ML (2012) The oil curse: how petroleum wealth shapes the development of nations. Princeton University Press, London
Ross ML (2014) What have we learned about the resource curse? Mimeograph. UCLA department of political science, Los Angeles, CA
Sachs DJ Warner AM (1995) Natural resource abundance and resource growth. Working paper 5398, National bureau of economic research, Cambridge, MA. Available at: http://www.nber.org/papers/w5398.Accessed 12 Feb 2016
Sachs DJ Warner AM (1997) Natural resource abundance and economic growth. Research paper, Centre for international development and Harvard Institute for international development, Cambridge, MA
Sala-i-Martin X Subramanian A (2003) Addressing the natural resource curse: an illustration from Nigeria. IMF working paper 03(139), International monetary fund, Washington, DC
Solow RM (1994) Perspective on growth theory. J Econ Perspect 8(1):45–54
Van der Ploeg F, Poelhekke S (2009) Volatility and the natural resource curse. Oxford Econ Pap 61(4):727–760

Chapter 4
Lessons from Other Petroleum-Rich States

Abstract The aim of this chapter is to assess the various policies adopted with respect to petroleum resource management in Norway, Kuwait, Azerbaijan and Nigeria. The author conducts a comparative analysis of the institutional design, human resource capacity development, and revenue management policy frameworks implemented by the four petroleum-producing countries in order to identify best practice. The main findings indicate that a petroleum resource curse is not inevitable. Prudent and sound institutional, human resource capacity building and petroleum revenue management policies help avoid transmission of the resource curse. This study suggests that a well-designed institutional governance model for the petroleum industry is vital to addressing common problems associated with effective economic and human resource development.

Keywords Resource curse · Institutional design · Human resource · Revenue management · Petroleum rich countries · Petroleum resource management

4.1 Introduction

The fact that some resource abundant countries, such as Norway, have escaped the resource curse while many others, such as Nigeria, have not is of particular interest to this study. The limitations of the economic literature on the resource curse lie in the inability of statistical analysis to answer a crucial question: why are some resource-rich countries able to utilise their natural resources to promote development while others are not?

The cross-country quantitative method of research is the most common analytical approach to analysis of the natural resource curse (Rosser 2006). However, as demonstrated in the previous chapter, this method falls short when attempting to explain obvious differences in the development experiences of various petroleum-rich countries. Ross (1999) highlights a gap between economic understanding and the politics of resource-rich countries. Furthermore, explaining causality in the transmission mechanisms of the resource curse requires deeper

K. Abdullah Auzer, *Institutional Design and Capacity to Enhance Effective Governance of Oil and Gas Wealth: The Case of Kurdistan Region*, Perspectives on Development in the Middle East and North Africa (MENA) Region, DOI 10.1007/978-981-10-4518-9_4

analysis of the particular natural resource at individual country level (Stevens and Dietsche 2008). As Maxwell (2004) emphasises, the qualitative approach considers causality in terms of processes and mechanisms, rather than simply demonstrating a correlation between existing variables. Ebbinghaus (2005) suggests that qualitative cross-country analysis is a more appropriate method than cross-country statistical analysis, based on assumptions regarding homogeneity, independence and the representativeness of the sample.

The qualitative method used in this chapter is a cross-country comparative approach (Bryman 2004; Rosser 2006). The goal is to demonstrate that, although statistical analysis of the resource curse seems to account for most of the difficulties confronting oil-rich states, a different picture is revealed when oil-abundant countries' economic, political and social policies and strategies are viewed in relative rather than absolute terms. From this perspective, the resource curse no longer appears to be inevitable or inescapable. Furthermore, undertaking cross-country comparative research will allow the Kurdistan region to learn from other countries' best practice in the governance of their oil and gas industries, enabling the Kurdistan Regional Government (KRG) to frame policy options for developing a long-term sustainable economy.

However, comparative analysis suffers from certain limitations, such as comparability, availability of datasets and generalisability. There is also the potential for differences, such as cultural factors, to be unobserved, while the historical, political and institutional context may make it difficult to replicate a policy from one country in another (Lijphart 1971; Synder 2001; George and Bennett 2005). Therefore, in this research an issue-oriented strategy (Stake 1995; Yin 2003) is used to analyse the adopted polices and measures across four selected cases. These policy areas are limited to the economic, political and socio-economic policies adopted by four major oil-exporting countries in response to the related transmission channels of the resource curse, based on the resource curse literature.

These countries—Norway, Kuwait, Azerbaijan and Nigeria—were selected on the basis of the research questions, the theoretical framework and prior knowledge of the cases (Ragin 1987). As major oil-exporting countries, considering their role in global energy security, the four case studies represent both the extremes and the middle range of oil-rich and resource curse characteristics. They also represent different approaches to governance of the oil and gas sector.

The following research questions framed the investigation of the best policy tools for addressing economic, political and socio-economic issues across the selected petroleum-producing countries:

1. How effective are institutional design and structure measures in addressing the political and economic issues associated with accountability, transparency and checks and balances?
2. How effective are policies and measures aimed at human resource capacity building?
3. How effective are economic policies in dealing with macroeconomic and petroleum revenue management issues?

Documentary sources were used for this cross-country comparative research. The general principles involved in handling and dealing with documentary resources are similar to those in other areas of social research. According to Scott (1990, p. 6), in handling documentary resources it is necessary to consider four criteria:

1. "Authenticity—whether the evidence is genuine and is of unquestionable origin;
2. Credibility—whether the evidence is free from error and distortion;
3. Representativeness—whether the evidence is typical of its kind and, if not, whether the extent of its specificity is known; and
4. Meaning—whether the evidence is clear and comprehensible."

Against this background, analysis of surveys, documents and reports published by governments relating to governance of the oil and gas sector provides information about the political, economic and social mechanisms that policy makers have chosen to manage their oil and gas resources and associated revenues.

The rest of this chapter is structured as follows. Section 4.2 starts with general information relating to the oil and gas industries of the selected countries, and then describes the implementation of institutional and administrative policies in governing the oil and gas sectors of the four petroleum exporters. Section 4.3 presents policies adopted by the respective governments to enhance their human resource capacity. Section 4.4 explains the economic policies adopted by the four cases to manage their petroleum revenues, as well as the measures taken to reduce their economic dependence on hydrocarbon revenues. Section 4.5 compares the institutional, human resource capacity-building and economic policies adopted by the four major oil-exporting countries, and Sect. 4.6 discusses the lessons learned from best practice.

4.2 Institutional Policy: The Role of Institutional Design and Frameworks in Petroleum Sector Governance

Institutional design should be taken into account in attempts to enhance the performance of the petroleum sector (Turber et al. 2011). Furthermore, setting up an effective petroleum-sector management system allows governments to make more efficient decisions on ways to manage their petroleum wealth.

The World Bank (Alba 2009) has set out an "extractive industry value chain" approach, which offers potential insights into the governance and institutional arrangements of petroleum-rich countries (Alba 2009). The natural resource management value chain focuses on institutional arrangements across the following key pillars of the extractive industry's value chain: (1) award of contracts or licences; (2) regulation and monitoring of operations; (3) revenue collection; and (4) revenue

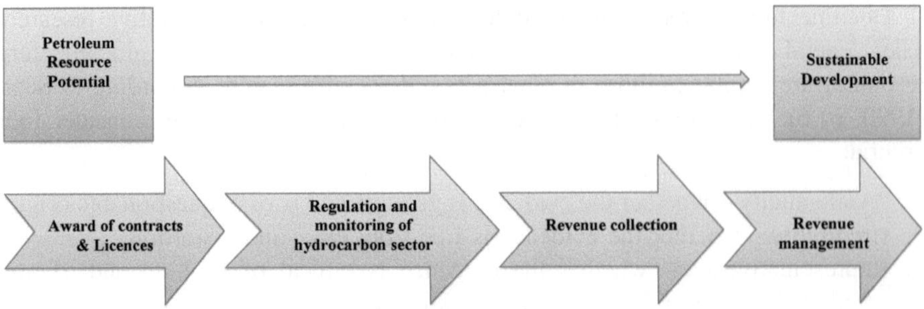

Fig. 4.1 Extractive industry value chain. *Source* Alba (2009)

management (Alba 2009, p. 3). This section considers the first three of these pillars, as shown in Fig. 4.1, while the fourth is discussed in Sect. 4.4.

4.2.1 Norway

Oil production began in Norway in 1971. The country has a population of more than five million. At that time, the country had well-functioning institutions, and shipping, fishing and the hydroelectric industry dominated its economy. However, petroleum exports formed 65% of its total merchandise exports in 2014 (World Bank 2015b).

Norway is now an important supplier of both oil and natural gas to other European countries. At the end 2014, its proven oil and natural gas reserves were about 6.5 billion barrels and 1.9 trillion m^3, respectively (BP 2015). All of Norway's oil reserves are located offshore on the Norwegian Continental Shelf (NCS), which is divided into the North Sea, Norwegian Sea and Barents Sea areas (NPD 2015). Norway is one of the major exporters of natural gas and crude oil. In 2013, it ranked as the third largest natural gas exporter after Russia and Qatar, and the twelfth largest exporter of crude oil in the world (OECD 2007; EIA 2014a). However, its petroleum production has gradually declined since 2001, as its oil fields have matured (see Fig. 4.2). The Norwegian Petroleum Directorate (NPD) recognises that maintaining production levels will be a big challenge for Norway as result of a continuing decline in crude oil production. Potential new discoveries will be the determining factor of future production levels (NPD 2015).

By contrast, the share of natural gas in Norway's total primary energy supply has grown since 1973, as it has progressively developed its large offshore gas fields on the NCS (IEA 2011). Only a small proportion of oil and natural gas is consumed domestically in Norway, as illustrated in Figs. 4.2 and 4.3. This is enabled by domestic reliance on renewable energies, including hydropower, wind and biomass (EIA 2014a). Consequently, Norway exports the vast majority of its oil and gas, and

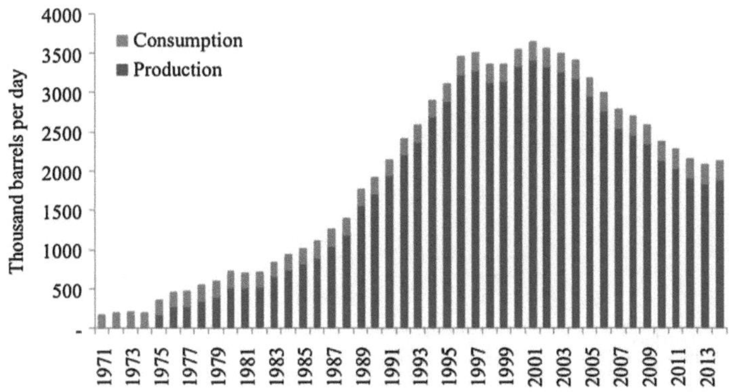

Fig. 4.2 Norway's crude oil production and consumption, 1971–2014. *Source* BP (2015)

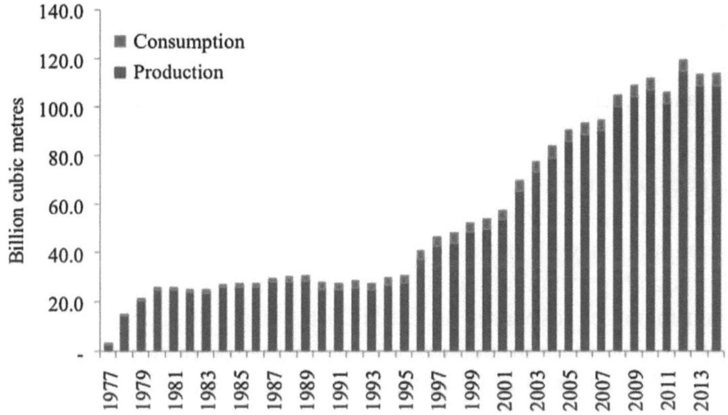

Fig. 4.3 Norway's natural gas production and consumption, 1977–2014. *Source* BP (2015)

is a significant exporter of crude oil and natural gas to EU countries, including the United Kingdom, Germany, France, the Netherlands, Belgium and Italy (NPD 2015). Norway is also a major supplier of refined fuel products, such as gasoline and diesel fuel, to European countries.

Figure 4.4 illustrates the location of the governance functions of the oil and gas sector in three separate organisational bodies: (1) policy-making (ministry), (2) regulation (NPD) and (3) operation (national oil company) (Hunter 2014).

Petroleum resources and all related revenues are the property of the Norwegian state. The fiscal regime in Norway is based on a concessionary system, under which the Norwegian government grants a licence to one or more oil companies giving exclusive rights for a limited period to the surveying, exploratory drilling and production of petroleum within the geographical area covered by the licence. This fiscal regime works through the payment of royalties and taxes to the state. Under the Petroleum Act, the Ministry of Petroleum and Energy (MPE) is authorised to

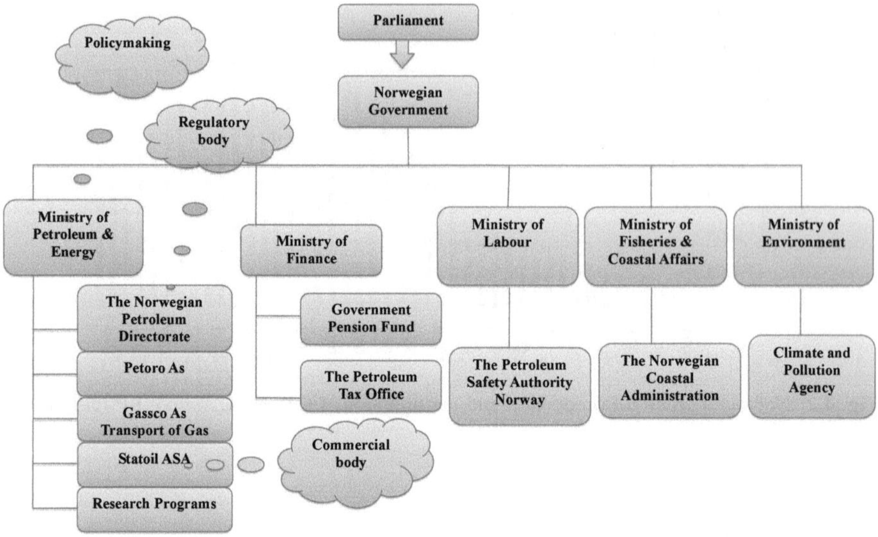

Fig. 4.4 Organisational structure of Norway's petroleum sector. *Source* NPD (2015, p. 17)

award licences and approvals for operators' field and pipeline development plans, while the NPD plays an advisory role in all steps of the licensing process when it comes to negotiation with individual companies (NPD 2015).

The Norwegian parliament has delegated responsibility for policy making to the MPE, which manages resources on the NCS to ensure that petroleum activities are carried out in line with guidelines set by the parliament and the government. The MPE is also charged with impact assessment regarding environmental and social aspects of petroleum activities prior to the opening of new areas. It fully owns two companies, Petro AS and Gassco AS, and partially owns Statoil. In order to gain control of the transmission of natural gas, the Norwegian government established a stated-owned company, Gassco AS, which manages the transportation of gas in the NCS and is the operator for Gassled (NPD 2015). The petroleum sector has cooperative links with a number of ministries, such as the Ministry of Labour, as well as the Ministry of Fisheries, the Ministry of Coastal Affairs and the Ministry of the Environment (NPD 2015).

The NDP is an independent technocratic agency and has a regulatory function. It reports to the MPE. Its key duty is to collect and analyse data relating to the NCS. When it was established in 1973, this was a challenging task due to its limited knowledge and resources (Engen 2009), but systematic efforts to increase its technical competence have transformed it into an independent, well-established regulatory body that monitors and controls petroleum activities in the Norwegian oil and gas sector. It is also an advisory body in the policy-making process (Engen 2009).

The Norwegian organisational model includes the national oil company (NOC), Statoil, which handles the commercial functions of the sector. It is a state-owned oil company established in 1972, with 50% participatory interest in all licences granted and a dominant role in the decision-making process. It also serves as a channel for technology transfer and economic development. Statoil has pursued the development of Norwegian industrial and technological capability through long-term research and development programmes, as well as through direct involvement in highly technological projects. It has also played a key role in the development of domestic service companies (Thurber and Istad 2010; Thurber et al. 2010b).

The Norwegian state restricted Statoil's power to a commercial function by transferring the state's holdings into the State Direct Financial Interest Agency (SDFI), which was established in 1985, eliminating its veto power over field decisions (Hunter 2014). Statoil also lost its right automatically to grant interest in all licences, in line with Article 4 of EEA agreement and EU Directive 94/22 EC, which requires objective, non-discriminatory granting of licences (Hunter 2014). Statoil was partially privatised in 2001. The Norwegian state has a 67% holding in Statoil, managed by the MPE. The major shareholders of the remaining 33% are the USA, Norwegian private owners, the rest of Europe, the UK, and the rest of the world (Statoil 2015).

The Norwegian state participates in production licences through a fully state-owned, non-operational company called Petoro. In addition to its commercial interest as a licensee, it regulates petroleum activities as a member of management committees in the field (Hunter 2014).

In summary, the primary goal of Norwegian petroleum governance has been to gain comprehensive state control over the petroleum industry; therefore, the NOC helped build domestic expertise through learning by doing. In this context, Statoil became a global competitive oil company. Norway's institutional design, which separates the different roles within the petroleum sector, has played a key role in increasing the checks and balances and avoiding conflicts of interest.

4.2.2 Kuwait

Oil production started in Kuwait in 1965. The country has a population of less than four million and has the fifth largest conventional oil reserves in the world. It is a small economy, with proven crude oil reserves of about 101.5 billion barrels and proven natural gas reserves of about 1.9 trillion m^3 at the end of 2014. Petroleum exports account for about 94% of total exports. Thus, its economy is heavily reliant on petroleum rents (EIA 2014b; BP 2015).

Kuwait's oil exports started in 1946 and it is now a major oil-exporting country. It is one of the few OPEC members with spare capacity, which is used to compensate when there is a shortage of supply in the world market. For example, it increased oil production in response to the loss of Libyan supplies in 2011 (EIA 2014b; BP 2015).

Production capacity is growing following a collapse in production resulting from the invasion of Kuwait by Iraq in 1990. The major customers for Kuwaiti oil are the United States, Europe, South Korea and India. As shown in Fig. 4.5, the proportion of domestic oil consumption is small: most of its production is available for export. However, the consumption of crude oil has been steadily increasing owing to the use of oil by power plants for electricity generation and by domestic refineries (EIA 2014b).

Kuwait has tended to develop its natural gas sector by attracting international oil companies to invest in the exploration and production of natural gas. Increasing domestic demand for natural gas and oil is attributable to the high consumption of combined power and water desalination plants. A subsidisation policy plays a major role in keeping electricity prices low, resulting in inefficient use. Therefore, in recent years the Kuwaiti government has stepped up efforts to increase its supply of natural gas to the domestic market. Kuwait's refinery sector grew during 2014, with a production capacity of 936 barrels per day, and Kuwait exports petroleum products to various global markets (EIA 2014b).

Kuwait is dependent on imported natural gas on account of its small natural gas production, which is unable to meet domestic demand (see Fig. 4.6).

Current fiscal policy for the petroleum sector is based on technical services agreements (TSAs). The Kuwaiti Constitution limits the involvement of international oil companies (IOCs) to the operation of local fields. TSAs provide a fixed rate of return on investment. Under such agreements, IOCs cannot acquire any interest in the underlying resource (Stevens 2008). Following nationalisation of Kuwait's petroleum industry in 1977, the entire oil and gas sector was brought under the control of the Supreme Petroleum Council (SPC), which is headed by the prime minister. The council is charged with setting government policy on petroleum wealth. However, it lacks the capability to address issues facing the Kuwait Petroleum Corporation (KPC), and hence has been unable to develop an effective strategy for the oil and gas sector. SPC is not accountable to parliament because the emir, rather than the parliament, appoints the government's cabinet. A lack of trust

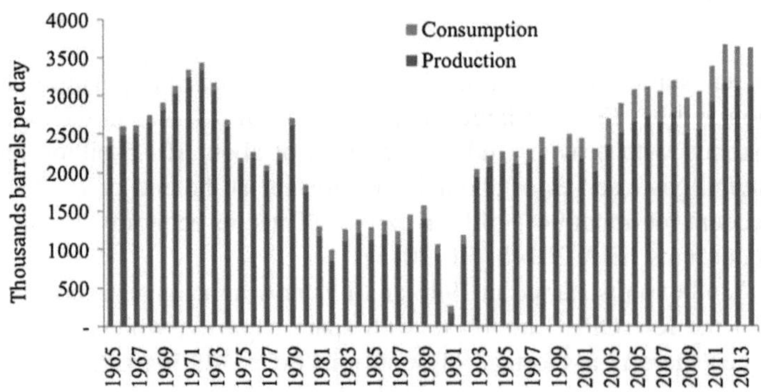

Fig. 4.5 Kuwait's crude oil production and consumption, 1965–2014. *Source* BP (2015)

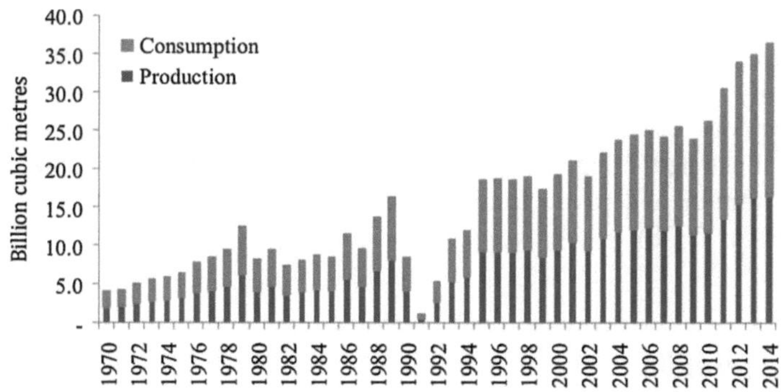

Fig. 4.6 Kuwait's natural gas production and consumption, 1970–2014. *Source* BP (2015)

between parliament and government has resulted in obstacles being set up by parliament to protect natural resources (Stevens 2008).

The Ministry of Oil is responsible for policy making and regulating the KPC. The Minister of Oil is the chair of KPC; thus, it is under the direct control of the Ministry, and has the power of veto over all KPC decisions. Consequently, the KPC seems not to be an independent entity in managing petroleum activities. The board of KPC comprises all directors of its subsidiaries. KPC is a fully state-owned company responsible for all petroleum activities. In 1980, the organisational structure of the petroleum sector was changed and KPC began to bring all related state-owned entities under a single umbrella, as shown in Fig. 4.7. KPC is accountable to its shareholder (the SPC) as well as parliament. It has faced difficulties in controlling its subsidiaries and implementing its projects effectively, a burden stemming from poor cooperation between the various subsidiaries and KPC,

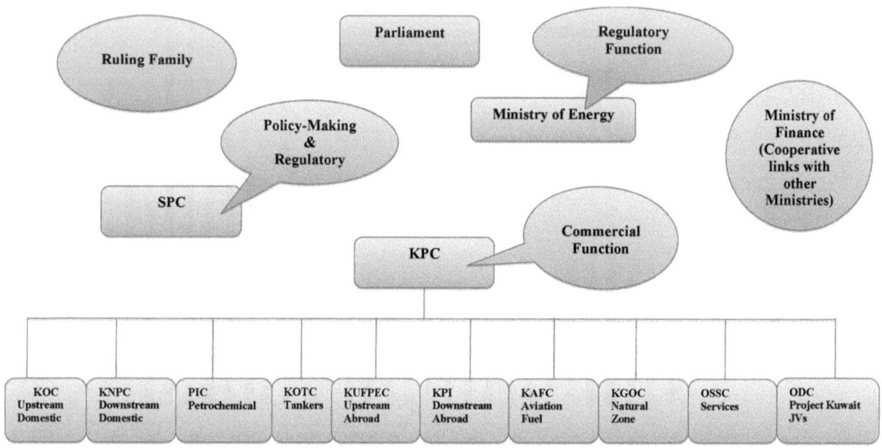

Fig. 4.7 Organisational structure of Kuwait's petroleum sector. *Source* Based on data from World Bank (2007), Stevens (2008)

long hierarchical and bureaucratic decision-making processes, political and elite interference, and low technical skills (World Bank 2007; Stevens 2008).

Authority over the oil sector is fragmented. Governance of the oil and gas sector lacks clarity of goals, roles and responsibilities between the agencies involved in the sector, as illustrated in Fig. 4.7. Moreover, in practice, there is no regulatory agency in the Kuwaiti oil and gas sector; thus, both the Ministry of Oil and PSC supervise KPC (Stevens 2008).

4.2.3 Azerbaijan

Azerbaijan is one of the world's oldest oil-producing countries. It is located in the South Caspian Sea basin with a population of more than nine million people. Since becoming independent from the Soviet Union in 1991, Azerbaijan's economy has relied heavily on its vast deepwater, offshore oil and gas resources in the Caspian Sea (EIA 2014c). British Petroleum (BP 2015) estimated Azerbaijan's crude oil and natural gas reserves at the end of 2014 to be 7.0 billion barrels and 1.2 trillion m^3, respectively. Crude oil exports account for 93% of total merchandise exports (World Bank 2015a).

As illustrated in Fig. 4.8, oil production in Azerbaijan grew from 307,000 barrels per day (bpd) in 2002 to 1.0 million bpd in 2010. However, the production volume has since declined, falling to 919,000 bpd in 2011 and 848,000 bpd in 2014. Falling production has contributed to technical problems in the Azeri-Chirag-Guneshi (ACG) field, which is Azerbaijan's main production field. Crude oil is exported through pipeline, truck and rail to world markets (EIA 2014c).

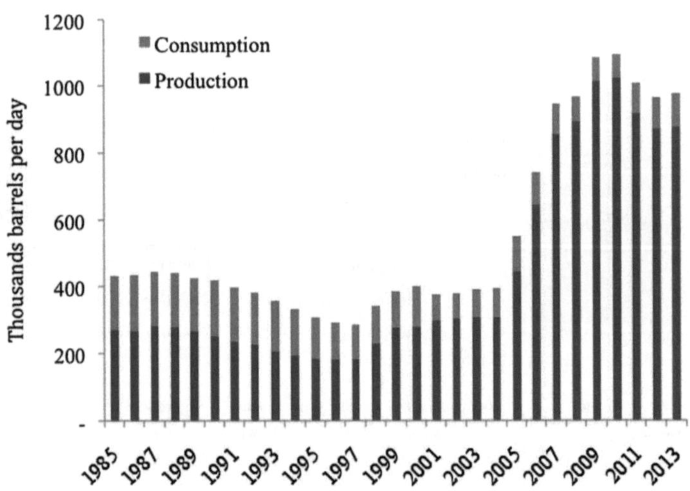

Fig. 4.8 Azerbaijan's crude oil production and consumption, 1985–2014. *Source* BP (2015)

Since 2007, natural gas production has increased with the commencement of production in the Shah Deniz field, developed by BP. Figure 4.9 also shows an increase in the consumption of natural gas, which is a major domestic energy source. The plan is that natural gas will be used for the country's future power generation. At present, the major natural gas customers are Turkey and Greece (EIA 2014c). Turkey is the major importer of Azerbaijani natural gas, and natural gas to Greece is transported through Turkey. However, under the 2011 Izmir agreement, Azerbaijan exports natural gas directly to Europe through Turkish territory. Azerbaijan has two refineries with a combined production capacity of 399,000 bpd as of 2014. The refineries need to be modernised to increase production (EIA 2014c).

In Azerbaijan, the fiscal regime is a contractual system. Under production sharing agreements (PSAs), Azerbaijan allows IOCs to invest in the oil and gas sector. The President of the Republic approves the rules for negotiations and contracts with IOCs, and the State Oil Company of the Azerbaijan Republic (SOCAR) is involved in the negotiation process. SOCAR has its own share in all contracts. At the same time, it acts as a government representative in contracts. This dual role results in conflicts of interest. Furthermore, SOCAR plays a major role in setting policy on petroleum resources as a result of its close cooperation with the president. It has a dominant role in the oil and gas industry, and thus it manages and regulates all petroleum activities. The role of the Ministry of Fuel and Energy is very limited (EIA 2014c; Energy Charter Secretariat 2011; Kjaernet 2010; Gojayev 2010).

In short, Azerbaijan has NOC-dominated governance model for its oil and gas sector. The NOC, namely SOCAR, has multiple responsibilities and functions, including policy-making, regulatory and commercial (see Fig. 4.10).

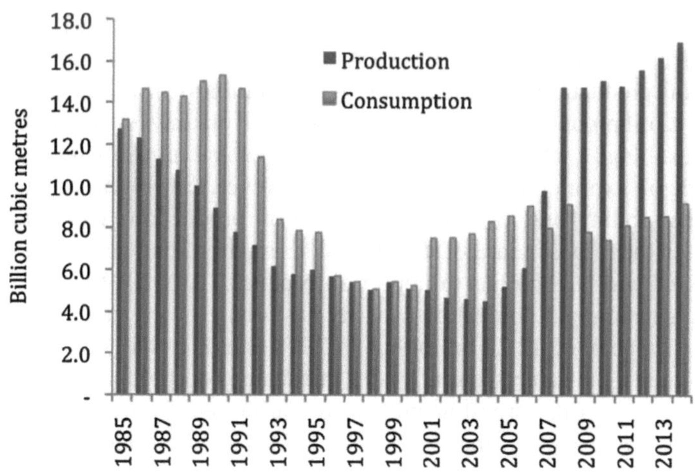

Fig. 4.9 Azerbaijan's natural gas production and consumption, 1985–2014. *Source* BP (2015)

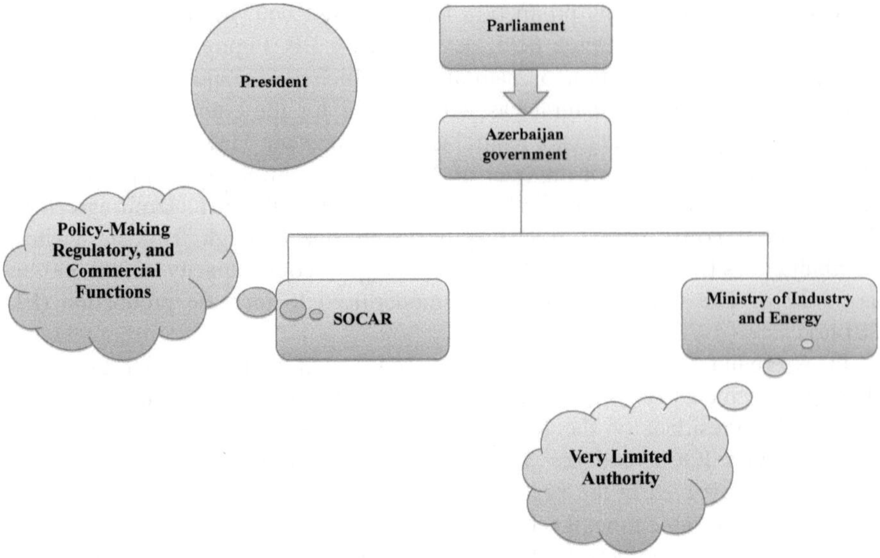

Fig. 4.10 Organisational structure of Azerbaijan's petroleum sector. *Source* Author, based on data from EIA (2014c), Energy Charter Secretariat (2011), Kjaernet (2010)

4.2.4 Nigeria

Nigeria is rich in oil and gas. It is located on the Gulf of Guinea on Africa's western coast, and has a population of 172 million. Nigeria is a member of the Organization of the Petroleum Exporting Countries (OPEC), and petroleum exports account for 88% of total national exports (OPEC 2014). Nigeria's petroleum reserve was estimated to be 37.1 thousand million barrels of crude oil and 5.1 trillion m^3 of natural gas at the end of 2014 (BP 2015).

Crude oil production reached a peak of 2.4 million bpd in 2005. As shown in Fig. 4.11, its oil production suffers from significant volatility, stemming from social conflicts resulting in attacks on oil pipelines and related infrastructure. Nigeria is a major crude oil exporter to the United States; however, the volume of exports to the US has been falling due to a growth in shale oil production. The refinery sector accounts for the largest share of domestic consumption, but capacity cannot meet local demand and Nigeria has to import oil products (EIA 2014d).

Nigeria's natural gas production capacity has been growing gradually since 2003 and reached a peak of 12.28 billion m^3 in 2008 (see Fig. 4.12), but this has been declining due to the same security issues threatening oil production capacity. It is the world's twenty-fifth largest natural gas producer, and the majority of its natural gas reserves are located in the Niger delta (EIA 2014d). Widespread natural gas flaring is a critical issue that the Nigerian government has failed to resolve. Despite the presence of a vast natural gas energy source, shortage of electricity remains an unresolved problem (Bloom 2010).

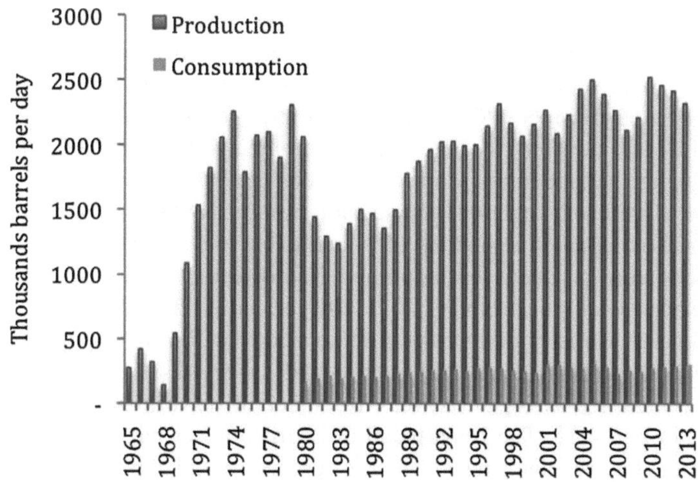

Fig. 4.11 Nigeria's crude oil production and consumption, 1965–2013. *Source* BP and EIA (2014d)

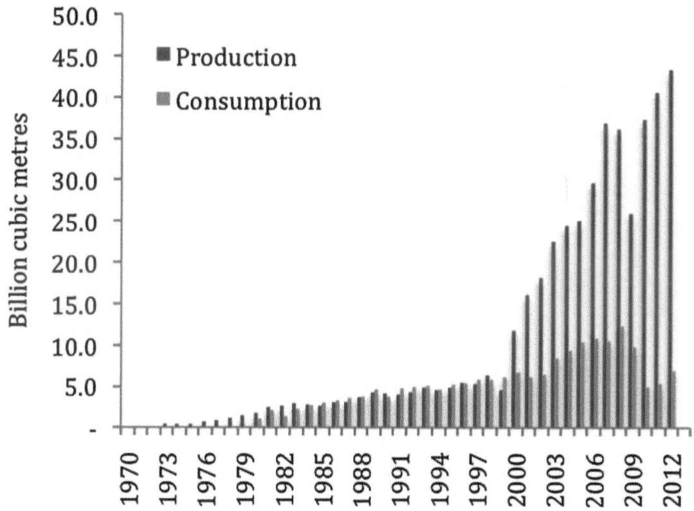

Fig. 4.12 Nigeria's natural gas production and consumption, 1965–2012. *Source* BP and EIA (2014d)

Nigeria has four refineries, which operate below full capacity because of operational failures, fires and sabotage, mainly on the crude pipelines feeding refineries. As a result, Nigeria has to import petroleum products. The government has planned for the construction of new refineries, but these have been delayed due to shortage of finance (EIA 2014d; OPEC 2014).

In view of the financial burden on the Nigerian National Petroleum Corporation (NNPC) in funding numerous joint venture operations, since the 1990s the Nigerian government has relied on production sharing agreements (PSAs) with IOCs (Thurber et al. 2010a). NNPC was created in 1977 to oversee regulation of the oil and natural gas industry. In 1988, the government divided NNPC into 12 subsidiaries, which expanded the scope of activities of NNPC in the oil and gas sector. The most significant functions of subsidiaries of NNPC include control over international companies operating in Nigeria, buying and selling crude oil and refined petroleum products, upstream and downstream operations, gas transportation and service activities (NNPC 2015; Thurber et al. 2010a).

The Federal Ministry of Petroleum Resources (FMPR) is responsible for initiating policies for the oil and gas sector, and supervises the implementation of approval policies. It is organised into a number of departments and agencies, as illustrated in Fig. 4.13. The Department of Petroleum Resources (DPR), within the FMPR, is the key regulator responsible for processing all applications for licences, for monitoring upstream and downstream activities, and for all rent payments in the oil and gas sector (NNPC 2015; Thurber et al. 2010a).

In summary, the governance model of the oil and gas sector in Nigeria has an overlapping institutional framework in which various agencies are involved. This results in duplications and conflicting regulatory functions.

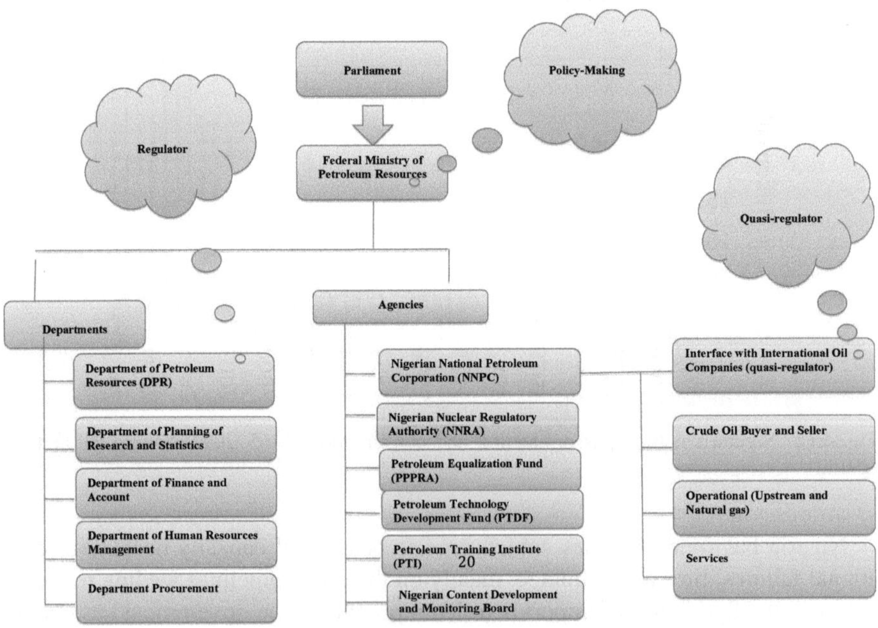

Fig. 4.13 Organisational structure of Nigeria's petroleum sector. *Source* Based on data from FMPR (2015), NNPC (2015), Thurber et al. (2010a)

4.3 Capacity-Building Policy: Role of Human Resource Development in Petroleum-Exporting Countries

Building capacity is a major determinant of the effective and efficient management of natural resources, particularly with regard to petroleum wealth. The shortage of local administrative and technical know-how, as well as new knowledge and technology, restricts the contribution of an indigenous labour force to industrial activities (ILO 2012; Marcel 2015). Support for education and training policies to build the right skills may contribute to economic prosperity through the improvement of productivity and growth. It contributes to social cohesion by increasing employment in good quality jobs and social engagement (OECD 2014). This section considers human capacity-building policies across the four cases.

4.3.1 Norway

The Norwegian government's policy for the oil and gas sector pursues the development of local capacity to maximise participation in exploration and development of the petroleum industry (Tordo et al. 2013). The establishment of a state-owned company stemmed from the need for an appropriate tool to implement national petroleum policy and also to fulfil the ambition of localising petroleum-related activities in order to increase national control over the country's oil and gas industry. Norway's advanced technological capacity in its shipbuilding industry facilitated the development of local engagement in the petroleum sector, which is located offshore (Wolf and Pollitt 2009; Heum 2008).

The MPE sought to enhance the technical capability and competiveness of NPD by providing incentives to attract adequate technical expertise (Thurber and Istad 2010). The MPE has also established various research programmes and partnerships covering a wide range of activities relating to employment creation, engagement of society in the petroleum industry, expansion of business markets, and building links with other industries (NPD 2015).

The Norwegian government's expenditure on education and training is 9% of GDP, which is the highest of all OECD countries (OECD 2014). With a vast fund for education and training purposes, the government aims to design a better-performing skills system in cooperation with employers, trade unions, universities, students and teachers. Furthermore, Norway's expenditure of GDP on research and development (R&D) is 1.61% of GDP. In addition to financial support, foreign companies operating on the NCS are committed to conducting at least 50% of research activities related to the development of fields in Norwegian institutions (Leskinen et al. 2012).

In summary, capacity building forms the key policy for the development of Norway's oil and gas sector.

4.3.2 Kuwait

Education expenditure accounts for 9% of Kuwait's GDP. However, the poor quality of the education and training system has led to a shortage of competent labour in Kuwait (IMF 2013). Kuwait has one of the lowest levels of expenditure on R&D among Gulf states, and there is no effective link between universities and industry (IMF 2014a).

Kuwaitis are employed mainly in the public sector. High pay and benefits for public jobs reduce incentives to work in the private sector, a classic problem in the "rentier states" of the Gulf (Gelb et al. 2002). The attractiveness of public employment also limits the number of workers engaging in entrepreneurship and skills training programmes to match their competence with the tradable labour market. Therefore, Kuwait suffers from a lack of specialised indigenous skills in the industrial sector (IMF 2013), a problem exacerbated by the small size of its population.

In addition, the lack of highly skilled workers is reflected in weak middle-level management across KPC's subsidiaries. The lack of managers with a deep knowledge of the oil industry stems from political interference in the recruitment process and the failure of KPC to devise a system that identifies training needs and promotes the most talented employees (Victor et al. 2011).

In short, the human resource capacity-building policy with respect to Kuwait's oil and gas sector has not been successful in educating and training the manpower to perform creditably the functions required by the petroleum industry.

4.3.3 Azerbaijan

There is growing concern about a decline in the number of participants in higher education institutions and the vocational training sector in Azerbaijan (Allahveranov and Huseynov 2013). Another factor that plays a major role in increasing the unemployment rate is a noticeable mismatch between the skills supplied by the educational system and labour market demands. Consequently, shortage of skills remains a significant barrier to productivity increases (Allahveranov and Huseynov 2013). In response, the government has adopted a range of measures to develop this capacity, such as introducing new curricula in academic institutions, new bachelor and master's programmes, and the establishment of new training centres (World Bank 2015a).

The Azerbaijani government uses local content rules in PSAs with foreign companies to develop its workforce through employment and training. These local content rules also target the development and procurement of local supplies and services. However, the effective implementation of regulations calling for local skills and suppliers is dependent on the quality and quantity of the domestic workforce and manufacturing companies (Tordo et al. 2013).

Human capacity-building for Azerbaijan's oil and gas industry faces many challenges, such as funding limitations and the low quality of the education system and training programmes.

4.3.4 Nigeria

The low skills of college graduates in Nigeria reflect the poor quality of the education system (Anyaehie and Areji 2015). This can be explained by a low level of public capital spending on human capacity development. Public funding of education and R&D is inadequate to meet the needs for development of the education system. The share of education spending in the 2013 budget was 8.5% (Barungi 2014). Consequently, the inadequacy of funding in education and training has a negative effect on economic development, thereby reducing living standards and welfare in Nigeria (Omoijmite 2011).

The local content policy introduced to develop a petroleum-competent Nigerian workforce started in 1971. In compliance with the Oil and Gas Industry Content Development Act of 2010, foreign petroleum companies in Nigeria must support the skills acquisition and empowerment of Nigeria's workforce and the enhancement of its existing manpower capacity. The local content rule also aims at effective involvement of Nigerian companies in the service and supply side of the oil and gas sector (Tordo et al. 2013).

The NNPC has established a Nigerian content development strategy investing in human capacity building. It awards joint venture agreements as a prime vehicle for the development of a range of skills in different disciplines, including engineering, geology and geophysics. However, achievement of the second objective of the content development strategy associated with enhancing the active participation of indigenous service companies in oil and gas activities is facing obstacles owing to poor fiscal policy and inadequate infrastructure (Balouga 2012).

The Nigerian government has set up a separate authority called the Nigerian Content Development and Monitoring Board that has responsibility for local content management. The Petroleum Technology Development Fund (PTDF) was established to develop domestic manpower for the oil and gas industry through training and education. For example, in order to fill the skills gap, overseas scholarships are granted to increase the number of master's and doctoral degree holders in several scientific disciplines related to the oil and gas industry (PTDF 2015).

The ineffective local capacity-building policy has resulted in an acute local skills shortage and a dearth in the competent technical manpower required by the oil and gas industry.

4.4 Economic Policy: To Maintain Macroeconomic Sustainability and Long-Term Economic Development

In the wake of the recent sharp fall in oil prices, as discussed in Chap. 2, economic growth and macroeconomic stability are big challenges facing oil-centred economies. According to the research reviewed in Chap. 2 and the results of Chap. 3, the uncertain and volatile nature of oil revenues, when they are the dominant source of government income, has a significant impact on inflation and the account balance in oil exporting countries and often delays sustainable growth of the non-oil economy. Stevens et al. (2015) emphasises the importance of an effective fiscal policy in response to unstable oil revenues in the major petroleum exporters. He argues that the more successful oil-exporting nations insulate their economy from fluctuating oil prices through the establishment of revenue stabilisation or savings funds. These well designed petroleum funds allow petroleum-exporting countries to manage their hydrocarbon revenues efficiently in order to stimulate long-term economic growth.

This section focuses on the fourth pillar of the extractive value chain, namely revenue management and long-term economic growth. The problems facing petroleum revenue management are two-fold. First, there is the question of large and unpredictable swings in oil prices, which make oil revenues uncertain. The second problem is associated with high dependence on the petroleum sector, which plays a dominant economic role in terms of generating national income. Therefore, it is important to analyse the economic policies and mechanisms that the four countries have undertaken to create a buffer against economically damaging oil shocks. It is also important to assess how they have formulated economic policies to ensure the diversification of the national income portfolio, and thus sustainable economic development in the post-petroleum era.

4.4.1 Norway

In the most successful petroleum-exporting countries such as Norway, the establishment of a sovereign wealth fund as a fiscal policy instrument serves both as a buffer for fluctuations in government revenues, and as a long-term investment. The Norwegian "Petroleum Fund" was set up in 1990 and was renamed the "Government Pension Fund" in 2006. The fund absorbs surpluses of petroleum revenues and pursues four objectives: (1) to shield the Norwegian economy from overheating and "Dutch disease"; (2) to save revenues for future generations; (3) to provide a fiscal buffer for "rainy days"; and (4) to ensure that physical oil and gas assets are gradually transferred into financial assets to replace oil and gas income as a source of state revenue (Ovesen 2008; Fearnley 2012).

The fund is integrated into the government budget, which means all spending and allocation should happen through the normal budget. A fundamental principle of Norwegian fiscal policy is a budgetary rule that restricts government spending to

4% of the balance of the fund over the course of an annual business cycle. Norges Bank manages the fund, and the Ministry of Finance sets the investment policy and the broad allocation of assets. Transfers from the fund require parliamentary approval (Humphreys et al. 2007).

The Government Pension Fund is divided into two parts: the Government Pension Fund Global (GPFG), which is invested abroad, and the Government Pension Fund Norway, invested in Norway. The GPFG holds 60% of its assets in equities, 35–40% in fixed income and as much as 5% in real estate. Its investments are spread globally outside Norway (Ovesen 2008; Fearnley 2012).

In addition to revenue management measures, Norwegian economic policies aim to reduce dependence on oil and gas revenues, thereby promoting Norway's long-term economic health and stability. Leskinen et al. (2012) highlight that the Norwegian local content policy aims for greater engagement of local service companies in the oil and gas sector. To achieve this goal, oil companies are obliged by law to provide the Ministry with a list of bidders. The Ministry is able to change the decision in favour of local companies if they are technically competitive. In this way, the Norwegian government plays an active role in ensuring that contracts are awarded to local companies by operators working in the Norwegian oil and gas sector. Heum (2008) explains another local content support mechanism adopted by the Norwegian government to enhance the quality of local suppliers through learning by doing: the government offers tax incentives to support foreign oil and gas companies working in local development programmes. Furthermore, the government aspires to make the Norwegian oil and gas industry globally competitive and able to contribute to economic growth, even if the domestic petroleum industry stagnates (Heum 2008). In addition, joint venture contracts between domestic and leading foreign companies have facilitated the transfer of technology, knowledge and expertise.

In this context, non-oil activities increased over the period 2000–2011; however, much of this growth was actually related to oil and gas production (IMF 2013). As shown in Fig. 4.14, oil rents fluctuated over the period 2000–2012 due to a gradual decline in petroleum production as a result of ageing oil fields. Since 2010, Norway's GDP has continued to grow despite a downward trend in oil and gas revenues. Thus, it can be concluded that Norway's dependence on the sale of hydrocarbons has not increased, owing to the presence of other productive sectors in its economic system such as the service sector, which contributes about 53% to GDP (Leskinen et al. 2012).

Since June 2014, the crude oil price has fallen by more than 50%, declining from over US$100 per barrel in January 2014 to around US$40 per barrel during the second part of 2015, the lowest since the bottom of the 2009 recession. This has reduced the return on investments associated with the Norwegian wealth fund, which stood at US$882 billion as of June 2015 (SWFI 2015b). The current decline in oil prices has increased unemployment in Norway, as the petroleum industry has reduced spending by laying off 10% of the sector's total workforce (Hovland 2014). Moreover, Statoil now has an international portfolio and is behaving more like an IOC in seeking cost cuts in the current low-price environment: "… Statoil needs an

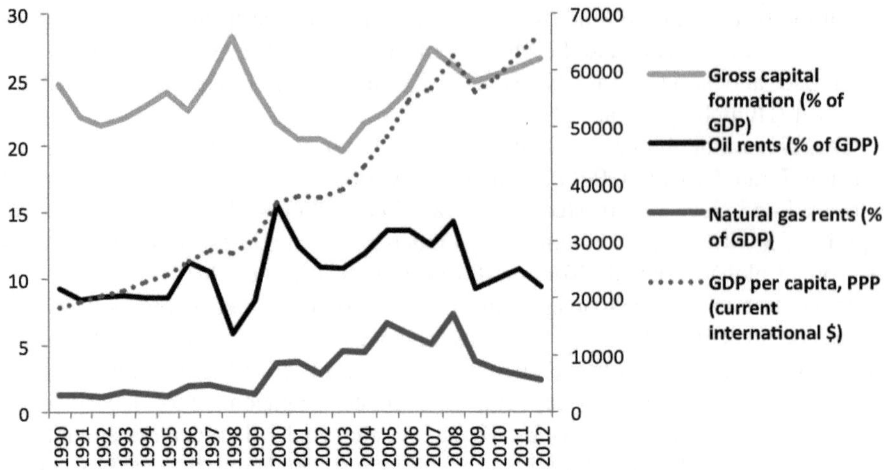

Fig. 4.14 Norway's economic indicators. *Source* World Indicator Development, World Bank (2015b). *Note* GDP per capita on the *right axis*; oil rents, natural gas rents, and gross capital formation on the *left axis*

oil price of between US$115 and US$120 this year to be able to pay its dividend without raising new debt" (Adams 2015).

In short, the Norwegian government's prudent oil investment policy has resulted in a decline in its economic dependence on the oil and gas industry. Oil rents have been used to diversify the Norwegian economy; however, the oil-based sectors are still subject to a downturn in oil prices.

4.4.2 Kuwait

In order to manage petroleum revenues, in 1953 Kuwait was the first country in the world to establish a sovereign wealth fund. The Kuwait Investment Authority (KIA) was established in 1982 and is now responsible for Kuwait's state assets. KIA manages two funds: the General Reserve Fund (GRF) and the Future Generations Fund (FGF). GRF was established in 1960 and is the main treasury for the state. All revenues, including petroleum rents, are transferred into GRF and all national expenditures are paid out of this fund. FGF was created in 1976 through a transfer of 50% of GRF's assets at that time. 10% of state revenues are now allocated to GRF annually. GRF's money is invested in assets, including bonds and cash asset classes, outside Kuwait. The purpose of KIA is to build up sufficient assets, and in turn achieve a long-term return on investment (KIA 2015).

The Kuwaiti government uses a range of means to distribute its oil rents. The major redistribution mechanisms of petroleum revenues include the following: (1) domestic investment in infrastructure and economic diversification programmes;

(2) land purchases and their sale at low prices to the public; (3) public transfer payments to Kuwaiti nationals and businesses; (4) subsidies, for example on electricity, water, food and housing; (5) public employment with well-paid salaries and benefit packages (91% of the Kuwaiti national labour force works in the public sector); (6) interventions in the private sector, including the provision of public infrastructure and services, with the exception of personal or corporate income tax; (7) regulation of Kuwait's FDI environment; and (8) involvement of Kuwaiti nationals in business by requiring foreign investors to sponsor them (El-Katiri et al. 2011).

In addition, Kuwait's economic policies pursue the goal of economic diversification through the establishment of a small and medium-sized enterprise fund to support industrial activities. 40% of industrial establishments supported by this fund are in trade, hotels and restaurants, and 33% are in construction and industry (IMF 2014a). The other major diversification is in the oil and gas sector, using crude oil and gas outputs to produce downstream products such as petrochemicals, fertilizers and chemicals, transport and logistics. Thus, Kuwait garners more value added from its endowments (Brinkley et al. 2012; El-Katiri et al. 2011).

Despite all these policies, Kuwait's economy is still overly-dependent on petroleum rents; hence, economic growth is subject to fluctuations in the global price of oil. As shown in Fig. 4.15, oil returns contribute more than 50% of GDP; therefore, a sustained decline in oil prices, and in turn oil rents, delivers a large, negative shock to the economy (Van der Ploeg and Poelhekke 2009). Furthermore, fiscal policy in terms of investment spending seems to be rather "procyclical" (IMF 2012) in Kuwait: investment expenditure increases after a positive rent shock and declines after adverse shocks in oil rents.

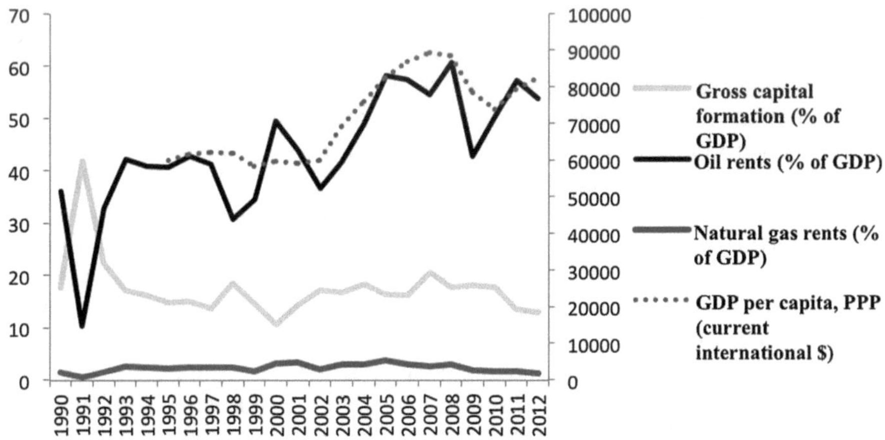

Fig. 4.15 Kuwait's economic indicators. *Source* World Indicator Development, World Bank (2015b). *Note* GDP per capita on the *right axis*; oil rents, natural gas rents, and gross capital formation on the *left axis*

The dramatic fall in oil prices has forced the Kuwaiti government to cut current spending. Restriction of subsidies, which continues at over 8–5% of total expenditure, has been the main fiscal policy response to the dramatic drop in current oil prices in Kuwait. However, it will not reduce capital spending on projects (Fattahova 2015). In addition, the large drop in oil prices has led to a government budget deficit. Consequently, the Kuwaiti state needs to balance its budget, which is based on oil valued at US$45 per barrel in 2015, down from its previous estimate of US$75 (Reuters 2015). The Kuwaiti sovereign wealth fund creates a buffer against the negative impact of oil price volatility, and Kuwait's petroleum fund is one of the largest in the world, valued at US$592 billion in June 2015 (SWFI 2015a).

4.4.3 Azerbaijan

The State Oil Fund of the Azerbaijan Republic (SOFAR) was established in 1999. The fund aims at macroeconomic stabilisation by smoothing public expenditure and growing non-oil sectors of the economy. It was set up to save revenues for the benefit of future generations. SOFAR is a legal entity managed by the Supervisory Council, and its membership is approved by the President of the Republic (SOFAR website 2014). The Azerbaijani sovereign wealth fund has been invested in a range of areas, such as real estate and gold (CESD 2012). SOFAR currently has US$37.3 billion in reserves (SWFI 2015a). Azerbaijan's high foreign exchange reserves serve as a buffer against oil price shocks.

The Azerbaijani government has sought to diversify its economy by setting priorities for budgetary expenditure in education, infrastructure and the development of non-oil sectors, but its economy still relies on the hydrocarbon sector. According to the Central Bank of Azerbaijan, the country's GDP growth in 2012 was driven entirely by public spending in non-oil sectors. Economic reform focuses on supporting small and medium-sized business establishments, developing the region and opening up the economy to private capital investment.

Azerbaijan's budget has become heavily dependent on petroleum revenues. The oil and gas industry provides more than half of the country's GDP, as shown in Fig. 4.16. Since 2003, annual GDP per capita rose as a result of increased oil and gas production, as did the value of exports because of the increase in oil prices. However, oil prices trended upward in 2011 and 2012, with the share of oil rents in GDP fluctuating because of a decline in oil production as a result of technical problems in the Azeri-Chirag-Guneshi (ACG) field. This decline in oil rents was not reflected in the GDP per capita due to growing public investment expenditure. This supports the evidence of Chap. 3 showing a decline in the impact of petroleum rents on economic growth due to increasing investment in non-oil projects.

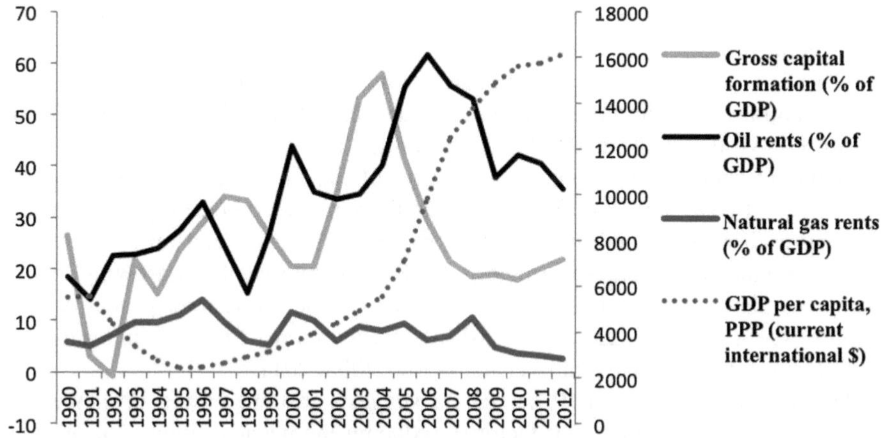

Fig. 4.16 Azerbaijan's economic indicators. *Source* World Indicator Development, World Bank (2015b). *Note* GDP per capita on the *right axis*; oil rents, natural gas rents, and gross capital formation on the *left axis*

The decline in oil prices in 2014 resulted in a big loss of budget revenue, and thus reduced the inflow of oil revenues into the State Oil Fund of the Republic of Azerbaijan (SOFAZ), which is the main source of transfers to the state budget. Transfers from SOFAZ into the state budget increased seven-fold over the period 2003–2013 (CESD 2012). In the wake of the oil price drop, increases in production volumes have been unable to limit losses in revenue for SOFAZ. Consequently, the Azerbaijani government has balanced its 2015 budget by setting the price of oil at US$90 per barrel, down from US$100 in 2014. In addition, low oil prices may lead to negative social consequences in terms of increased unemployment, as SOCAR and BP have reduced their costs by cutting 8% of their workforce (Farchy 2015).

4.4.4 Nigeria

In recent years, Nigeria has taken several initiatives to set up a sovereign wealth fund. The first fund, the Excess Crude Account (ECA), was established in 2004 and aims to save surplus oil revenues according to a conservative benchmark oil price, thereby insulating the budget and the economy from fluctuating oil prices. The second fund is the Nigeria Sovereign Investment Authority (NSIA), established in 2011 following parliamentary approval. This fund serves as an instrument to support sustainable development strategies and thus long-term economic growth (Brown et al. 2014).

NSIA is divided into three further funds based on its multiple policy objectives: the Nigeria Infrastructure Fund, the Stabilisation Fund, and the Future Generation Fund (NSIA website 2015). The Finance Ministry is responsible for overseeing these accounts. Most of the excess oil revenues allocated to the ECA have already been used (Wallis 2014), and it has US$2 billion in 2015 (IMF 2015).

While oil prices have become increasingly volatile since the 2000s, oil rents as a percentage of GDP have also fluctuated over the same period (see Fig. 4.17) due to production disruptions and the activities of crude oil thieves and oil pipeline vandals, as explained earlier. Nevertheless, GDP per capita has been growing, owing mainly to the continuing positive performance of non-oil sectors, primarily agriculture, services and trade (IMF 2014b). Furthermore, Fig. 5.1 indicates that investment spending has risen over the period 2008–2011.

This upward trend in GDP per capita suggests that increasing investment in non-oil projects is an important factor in reducing the negative impact of oil rents on economic growth. This finding again supports the results of Chap. 3. Furthermore, the Nigerian government tend to promote manufacturing sector by encouraging small and medium scale enterprises, which can play a major role in creation job opportunities, skill development and poverty alleviation (Aigboduwa and Oismaoje 2013).

In the wake of the recent fall in oil prices, the Nigerian government has revised its 2015 state budget based on an oil price of US$65 per barrel, rather than the previous assumption of US$77.5 per barrel (Hou et al. 2015). In addition, the Nigerian government intends to double its value added tax income, reduce public investment expenditure by delaying or cancelling projects (Mcgrorarty et al. 2015), and reduce fuel subsidies (Hou et al. 2015).

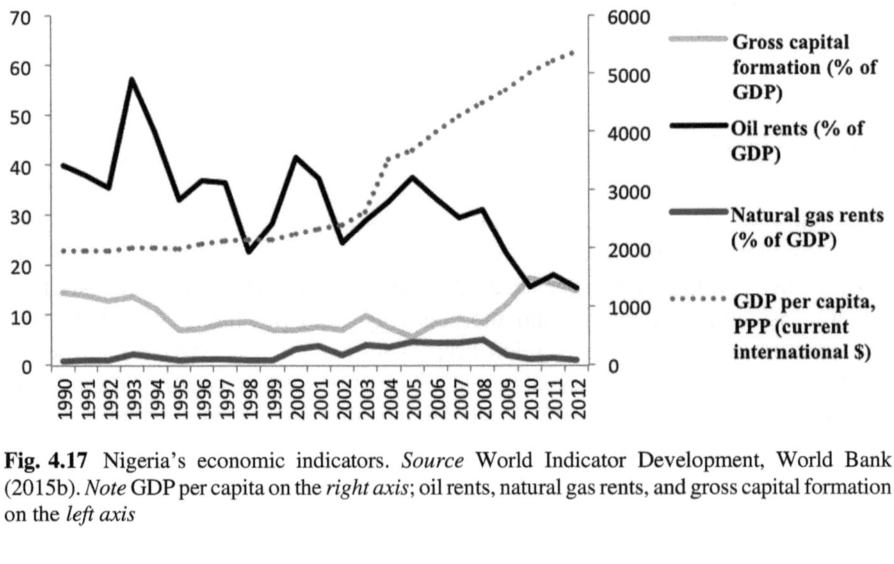

Fig. 4.17 Nigeria's economic indicators. *Source* World Indicator Development, World Bank (2015b). *Note* GDP per capita on the *right axis*; oil rents, natural gas rents, and gross capital formation on the *left axis*

4.5 Assessment

The previous sections have discussed the various policies adopted with respect to petroleum resource management in Norway, Kuwait, Azerbaijan and Nigeria. This section presents a comparative analysis of the institutional design, human resource capacity development, and revenue management policy frameworks implemented by the four petroleum-producing countries in order to identify best practice.

Table 4.1 provides a comparison of the four variables used in revenue management policy between the case studies. Moreover, it indicates that the fuel rent share is negatively associated with gross capital formation. It also shows that the countries with the oil sector as a dominant economic sector have spent less of their oil revenues on capital investment.

4.5.1 Institutional Design

With regard to the institutional design of petroleum sector governance, Norway has opted for an oil institution model based on a separation of powers between the

Table 4.1 Descriptive Statistics

	Mean	Standard deviation	GDP per worker	Gross capital formation	Oil rent	Natural gas rent
Norway						
GDP per worker	39,156.42	16,020.35	1			
Gross capital formation	23.65	2.30	0.41	1		
Oil rent	10.58	2.35	0.48	−0.21	1	
Natural gas rent	3.13	1.86	0.69	0.02	0.80	1
Kuwait						
GDP per worker	71,580.57	11,477.66	1			
Gross capital formation	17.45	5.89	0.42	1		
Oil rent	43.74	11.31	0.83	−0.01	1	
Natural gas rent	2.29	0.77	0.06	0.06	0.25	1
Azerbaijan						
GDP per worker	7093.03	5006.58	1			
Gross capital formation	25.69	13.29	−0.18	1		
Oil rent	34.78	13.20	0.55	0.24	1	
Natural gas rent	7.74	2.95	−0.60	0.17	0.06	1
Nigeria						
GDP per worker	3057.83	1240.90	1			
Gross capital formation	10.218	3.45	0.36	1		
Oil rent	32.28	9.95	−0.67	−0.23	1	
Natural gas rent	2.29	1.46	0.30	−0.46	0.14	1

NOC, which is engaged in commercial hydrocarbon operations, the regulatory body, which provides oversight and technical expertise, and the government ministry, which helps set policy.

Statoil has sought to enhance its commercial competiveness and operational performance, thereby increasing its financial returns to the state, by carrying out extensive oil and gas operations both in Norway and abroad. Furthermore, exploration and production agreements between Statoil and foreign companies are designed to increase Norwegian state control over the petroleum industry by strengthening Norwegian expertise in exploration and production.

The Norwegian Petroleum Directorate has advisory and regulatory functions, and parliament is responsible for the overall framework. These well-designed institutions are underlying factors that prevent any conflict of interest and have contributed to successful policy implementation in Norway's petroleum industry. Norway's experience with its oil institutional structure is different from that of the other cases here, with a strong emphasis on transparency and accountability.

Conversely, Kuwait's oil and gas industry lacks a clear administrative body. The parliament does not appoint the government, and only has the ability to veto rather than initiate major policy decisions; hence, many petroleum decisions have been halted by parliamentary veto. Through veto power, the oil minister, as chairman of KPC's board, monitors the extent of its control and authority. KPC suffers from inadequate authority to initiate and deliver needed political and regulatory reforms.

Taking into account the very poor state of Kuwait's economy prior to the discovery of oil and its lack of expertise in exploration, the concessionary fiscal system was the best option for its government. Following nationalisation of its petroleum wealth, the state-owned oil company took control of oil and gas activities, while the participation of foreign petroleum companies was limited to service petroleum contracts.

However, this strategy has had to change due to a pressing need for technical assistance from IOCs to work on the country's new fields. KPC lacks experienced management and has insufficient technical knowledge as a result of political and personal interference in the appointment and employment of personnel.

In Azerbaijan, the Ministry of Industry and Energy is responsible for the preparation and implementation of state policy in the oil and gas sector. In contrast to Norway and Kuwait, Azerbaijan's oil ministry plays both a policy-making and a regulatory role in the oil and gas industry. SOCAR is a state-owned oil company and acts as the commercial body of the oil and gas industry. It also has its own (state) share in all contracts and is thus a contractor.

At the same time, SOCAR represents the government in PSA contracts. As a contractor, SOCAR is interested in increasing its share in the contract; however, this might also reduce government revenues. Consequently, a conflict of interest has come about. This has occurred partly due to a lack of adequate law and regulation over the oil industry, and partly due to the considerable power of SOCAR and its influence over the oil and gas sector (Ciarreta and Nasirov 2012).

Nigeria's oil and gas sector has an overly complicated and inefficient administrative design, resulting in the duplication of functions between NNPC and DPR. NNPC is a commercial entity, but is also a quasi-regulator with very diverse activities. Its operations are more focused on oil marketing and downstream functions. Nigeria's oil and gas sector is negatively affected by an ineffective bureaucratic system, as in the hydrocarbon sector in Kuwait and Azerbaijan.

Nigeria now uses PSAs rather than the joint venture model, which historically was the main form of agreement between NNPC and IOCs. As with Azerbaijan, the technical inability of the state-owned oil company (NNPC) and funding problems are reasons behind the use of PSAs. NNPC and SOCAR are ineffective in monitoring the operational activities of foreign oil companies; thus, contracts are weakly enforced. For example, Watts (2004) maintains that oil companies have not made serious efforts to report the results of environmental assessments of their extractive activities in Azerbaijan and Nigeria.

4.5.2 Developing Human Resource Capital

A review of the human resource capacity-building policies of the four selected countries highlights that Norway benefits from a competent labour force, thanks to its efficient, transparent and accountable governance structure and institutions. The terms of IOCs' licences make it mandatory to transfer skills and competence to Norwegian companies. Norwegian oil companies, mainly Statoil, set personnel training as a key objective. In addition to training schemes, the transfer of technology and the development of research by promoting cooperation between IOCs and Norwegian research institutions is one of the most successful aspects of Norway's hydrocarbon policy.

Unlike Norway, the other countries are suffering from the limited technical competence of their state-owned companies and a shortage of research activities in the oil and gas sector. Kuwait lacks skilled labour at competitive prices, as well as competent management. It has historically imported most of its workers, particularly those with high skills. Similarly, low human resource and limited research capacity are serious challenges faced by Azerbaijan and Nigeria. The NOCs in these three countries are very ineffective and rely on foreign companies to perform the most complex functions in their oil and gas sectors.

4.5.3 Revenue Management

In terms of revenue management, all of the countries under investigation have adopted institutional vehicles to combat the negative effects of fluctuating oil revenues. In contrast to the other cases, Norway's sovereign wealth fund has favoured policies aimed at sound oil revenue management, including: (1) a strict and precise fiscal rule that constrains transfers to the non-oil deficit budget; (2) integration into the overall fiscal framework; (3) approval of spending decisions by parliament; and (4) partial investment abroad. Norway has diversified its sources of income, partly by investing its petroleum revenues abroad in order to diversify risk and increase the expected rate of return. This diversified investment portfolio insulates the domestic economy from Dutch disease and external shocks. The non-oil tradable sectors are gradually growing; however, the greater proportion of non-oil tradable sectors is still linked to hydrocarbon industries.

Unlike Norway, the governments of Kuwait, Azerbaijan and Nigeria have been unwilling to impose fiscal rules on their respective funds. A lack of firm fiscal rules regarding inflows and outflows, as well as inadequate independent oversight, has resulted in arbitrary withdrawals of funds. In addition, oil funds have not been integrated into the state budgetary process, and powerful political and elite groups control transactions from the fund. Given this context, in contrast to Norway, transparency and accountability remain a major challenge for Kuwait, Azerbaijan and Nigeria (Bauer 2014).

One explanation for the major differences between the selected countries lies in their contrasting economic backgrounds. Since the early 1970s, Norway has had a developed and diversified economy, with a highly educated population and a good standard of living. Its economy does not rely directly on petroleum revenues. This is in contrast to Kuwait, Azerbaijan and Nigeria, which depend heavily on oil and gas rents. However, economic reforms are being introduced through increases in public capital expenditure, leading to the growth of non-oil sectors. The insufficient number and low quality of local graduates, poor infrastructure and ineffective investment regulations to attract foreign investors are major factors that have led to declining productivity and competitiveness in the petroleum and non-petroleum sectors of Kuwait, Azerbaijan and Nigeria.

A comparison between Norway's GDP per capita and those of the other three countries shows that Norway's rate seems relatively stable, while those of Kuwait, Azerbaijan and Nigeria indicate a greater correlation with the negative and positive oil price shocks of the 1990s and 2000s, as illustrated in Fig. 4.18. This implies that the high proportion of petroleum revenues contributing to GDP makes the economy vulnerable to changes in the price of the dominant commodity, on account of limited non-oil exports from Kuwait, Azerbaijan and Nigeria.

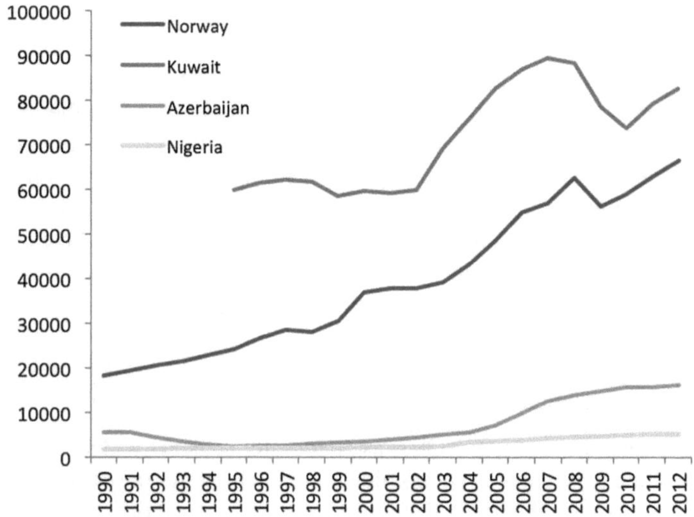

Fig. 4.18 GDP per capita (PPP, current international $), 1990–2012. *Source* World Indicator Development, World Bank (2015b)

4.6 Conclusions

This assessment of the institutional policies of the four selected petroleum exporters illustrates the importance of institutional design in the oil and gas sector as a determinant of the effective governance of the petroleum industry. The design of institutions should establish the goals, roles and responsibilities of the main stakeholders involved in the oil and gas sector. The key actors are the ministry, including an independent, highly capable regulatory agency, foreign and national oil firms, and parliament (Lahn et al. 2009). The development of such an administrative structure provides a sound business environment, reduces excessive bureaucracy and provides stability. Indeed, clarity of goals, roles and responsibilities between agencies or departments is crucial in mitigating the opportunity for patronage activities, and increases levels of transparency, accountability and checks and balances.

Analysis of policies associated with developing indigenous technological capacity shows that, with the exception of Norway, the respective governments have been unable to support local skills development, capacity building and utilisation. A lack of highly skilled manpower has served to increase technical dependence on foreign oil companies for oil and gas development in Kuwait, Azerbaijan and Nigeria. The dearth of effective policies to develop higher education systems or cooperation and links between local universities and research institutions has led to the poor performance of the domestic petroleum industries (Glyfason 2001).

The oil and gas sector is the major source of government revenue in Kuwait, Azerbaijan and Nigeria. However, all of these countries have attempted to diversify their economies through the effective allocation of oil funds to diverse capital investment areas. Kuwait's diversification policy concentrates on the expansion of horizontal links within the petroleum industry, which still rely on oil and gas, while in Azerbaijan and Nigeria public capital spending has been a driver of non-oil growth, including the construction and service sectors. An improved business environment is crucial to enabling increased private investment. There is a need for more effective public spending on infrastructure in both Azerbaijan and Nigeria. However, the creation of non-petroleum-based links still seems to be a challenge for all selected countries.

Several key lessons emerge from this comparative analysis that may be relevant to the effective governance of oil and gas in the Kurdistan region:

(1) Governments have used NOCs as a tool to achieve wider socio-economic policy objectives, such as economic diversification and raised local educational levels. However, the dual functions of NOCs, as both regulators and oil companies, may discourage IOCs from investing in the oil and gas sector. Furthermore, the regulatory function of NOCs impacts negatively on the efficiency of projects, and a potential conflict of interest emerges when an NOC acts simultaneously as a government representative and in pursuit of commercial goals. An independent NOC board may reduce political interference in the decision-making process and delays in the approval process, as well as enabling the NOC to maximise its limited technical capacity and become more transparent and accountable.

(2) A government ministry or independent entity should carry out the functions of regulation and oversight. However, the technical capability of the regulatory body is important for the effective monitoring and control of all hydrocarbon activities, thus ensuring a comprehensive follow-up of petroleum activities. The regulatory agency may also provide technical assistance to policy makers and thus act as an advisory arm.

(3) Efficient local content measures are crucial for increasing levels of productivity and competitiveness in oil- and gas-exporting countries. This key strategy pursues two major goals: localisation of the workforce and the development of local servicing companies. Furthermore, mechanisms for enhancing the quality of the education and vocational systems, staff training and a constant upgrading of workers' skills are the pillars of effective local content policy.

(4) A well-functioning petroleum fund is an important element in the process of moving away from a hydrocarbon-based economy. An economy is well served by a petroleum fund when it is run according to precise fiscal rules and is integrated into the state budget. In such a context, funds can be immunised against political interference. The petroleum fund acts as a catalyst for the creation of economic links with the wider economy. Petrochemical manufacturing and service sectors account for the creation of economic links within the oil and gas sectors: the service sector plays a major role in increasing economic

Table 4.2 Summary of comparative cross-country case studies

Characteristics	Norway	Kuwait (OPEC)	Azerbaijan	Nigeria (OPEC)	Notes
General information					
Geography	Europe	Middle East	Caspian	Africa	
Access to sea	Yes	Yes	No	Yes	
Reserves at end 2014					
Oil (billion barrels)	6.5	101.5	7.0	37.1	
Gas (trillion m^3)	1.9	1.8	1.2	5.1	
Production at end 2014					
Oil (thousand bpd)	1657(exports)	2618 (exports)	747 (exports)	2361 (total production)	
Gas (billion m^3 daily)	104.1 (exports)	3.7 (imports)	7.7 (exports)	38.6 (total production)	
(Refinery capacity (thousand bpd)	316	936	–	–	
National oil company	Statoil	KPC	SOCAR	NNPC	
Fuel export (% of merchandise exports) (%)	67	94	93	88	
Governance of oil and gas sector					
Fiscal regime	Concessionary system: licence	TSA	PSA	Joint venture and PSA	Technical and financial capabilities play major role in selection of model for exploration and production agreement
Policy-making function: set policy of oil and gas sector	Ministry	PSC	Ministry	Ministry	In Norway, policy-making body is separate from regulatory body, but not in Azerbaijan and Nigeria. Kuwait has an extra agency for policy-making function

(continued)

Table 4.2 (continued)

Characteristics	Norway	Kuwait (OPEC)	Azerbaijan	Nigeria (OPEC)	Notes
Regulatory function: monitor and oversee petroleum activities	NPD	Ministry	SOCAR and ministry	NNPC and ministry	In Nigeria different agencies are involved in regulatory functions Technical competence of regulatory body is to effectively curtail and oversee all activities in oil and gas sector
Operator function: Role and degree of autonomy of NOC	NOC: Statoil has high technical capacity and plays major role in technical transfer and development of domestic supply sector Also IOCs	KPC	SOCAR	NNPC	In Norway Statoil has commercial function and is partially privatised In Nigeria, because NNPC acts as regulatory body and in PSA contracts as NOC, multiple functions cause conflicts of interest Same applies to Azerbaijan due to multiple roles of SOCAR
Human capacity building					
Education	High quality and adequate funding	Low quality and low funding	Low quality and low funding	Low quality and low funding	Common issue is mismatch of supply and market demand
Training system	NOC plays key role Well-designed programme and R&D	NOC plays poor role Inefficiently functioning programme and R&D	NOC plays poor role Inefficiently functioning programme and R&D	NOC plays poor role Inefficiently functioning programme and R&D	Norway introduced effective regulation and funding to develop local capacity Kuwait relies on expatriates and ineffective training system
Cooperation with research and academic institutions	Establishment of petroleum research centre and cooperation with local and international universities and research institutions	Low funding and cooperation with research institutions	Low funding and cooperation with research institutions	Low funding and cooperation with research institutions	Norway has introduced regulation obliging IOCs to conduct research in Norway

(continued)

Table 4.2 (continued)

Characteristics	Norway	Kuwait (OPEC)	Azerbaijan	Nigeria (OPEC)	Notes
Petroleum revenue management and economic diversification					
Macroeconomic stabilisation	Saving and future generation fund	Saving and future generation fund	Saving fund	Infrastructure, saving and future generation fund	Design and fiscal rules of the fund plays a major role in effective and transparent use of petroleum revenues
Diversification	Growth of non-oil sectors Within petroleum sector and other non-oil sectors	Low growth of non-oil sectors More within petroleum sector	Low growth of non-oil sectors Reform consists of increase in public capital spending, such as infrastructure and services	Low growth of non-oil sectors Reform consists of an increase in public capital spending, such as infrastructure and services	Norway's economy is less dependent on petroleum sector, thanks to its non-oil industries Other countries tend to increase productive activities to diversify sources of revenue generation by increasing capital spending. This is reflected in growth of GDP in recent years Lack of adequate infrastructure development

productivity and, in the context of local content policy, local service companies accelerate the transfer of skills and technologies and the creation of jobs. In addition to the hydrocarbon-related sectors, other non-petroleum productive sectors, such as agriculture and industrial manufacturing, appear to be determinants of the success or failure of petroleum-abundant states in the development of sustainable economic growth.

Best practice in petroleum wealth management indicates that a petroleum resource curse is not inevitable. Prudent and sound institutional, human resource capacity building and petroleum revenue management policies help avoid transmission of the resource curse. This study suggests that a well-designed institutional governance model for the petroleum industry is vital to addressing common problems associated with effective economic and human resource development. The key points of comparative cross-country case studies are summarised in Table 4.2.

References

Adams C (2015) Statoil to cut spending after oil price collapse. *The Financial Times*, 22 February. Available at: http://www.ft.com/cms/s/0/f693d4b4-adcb-11e4-8188-00144f. Accessed 12 Feb 2016

Aigboduwa JE, Oisamoje MD (2013) Promoting small and medium enterprises in the Nigerian oil and gas industry. Eur Sci J 19(1):244–261

Alba EM (2009) Extractive industries value chain: a comprehensive integrated approach to developing extractive industries. Africa Region Working Paper Series No. 125, World Bank, Washington, DC

Allahveranov A, Huseynov E (2013) Costs and benefits of labour mobility between the eu and the eastern partnership partner countries—country report: Azerbaijan. European Commission, Geneva, Switzerland

Anyaehie MC, Areji AC (2015) Economic diversification for sustainable development in Nigeria. Open J Polit Sci 5(2):87–94

Balouga J (2012) Nigerian local content: challenges and prospects. Int Assoc Energy Econ 6 (3):23–26

Barungi B (2014) Nigeria 2014. Issy les Moulineaux. African economic outlook, France. Available at: http://www.africaneconomicoutlook.org/en/country-notes/west-africa/nigeria/. Accessed 11 Feb 2016

Bauer A (2014) Managing the public trust: how to make natural resource funds work for citizens. Natural Resource Governance Institute, New York. Available at: http://www.resourcegovernance.org/sites/default/files/NRF_Complete_Report_EN.pdf. Accessed 11 Feb 2016

Bloom DE (2010) Nigeria: the next generation report. Working paper 6210, Program on the Global demography of aging. Harvard University, Cambridge MA

BP (2015) BP statistical review of world energy, 64th edn. Available at: http://www.bp.com/content/dam/bp/pdf/energy-economics/statistical-review-2015/bp-statistical-review-of-world-energy-2015-full-report.pdf. Accessed 11 Feb 2016

Brinkley I, Hutton W, Schneider P, Coates Ulrichsen K (2012) Kuwait and the knowledge economy, Report 22, The work foundation and the Kuwait programme on development, governance and globalisation in the Gulf States, London school of economics, London

Brown BA, Naenwi ACM, Dimoji MO, Onwunem FA, Lovina O (2014) Excess crude account and sovereign wealth fund as strategic tools for sustainable development in Nigeria. J Econ Sustain Dev 5(2):57–61

Bryman A (2004) Social research methods, 2nd edn. Oxford University Press, Oxford

CESD (2012) SOFAZ's tardy investment policy. Center for economic and social development website commentary. Available at: http://cesd.az/new/?p=4382. Accessed 11 Feb 2016

Ciarreta A, Nasirov S (2012) Development trends in the Azerbaijan oil and gas sector: achievements and challenges. Energy Policy 40:282–292

Ebbinghaus B (2005) When less is more: selection problems in large-N and small-N cross-national comparisons. Int Sociol 20(2):133–152

EIA (2014a) Norway. US Energy Information Administration, Washington, DC. Available at: https://www.eia.gov/beta/international/analysis.cfm?iso=NOR. Accessed 12 Feb 2016

EIA (2014b) Kuwait. US Energy Information Administration, Washington, DC. Available at: http://www.eia.gov/beta/international/?fips=KU. Accessed 12 Feb 2016

EIA (2014c) Azerbaijan. US Energy Information Administration, Washington, DC. Available at: http://www.eia.gov/beta/international/?fips=AJ. Accessed 12 Feb 2016

EIA (2014d) Nigeria. US Energy Information Administration, Washington, DC. Available at: http://www.eia.gov/beta/international/?fips=NI. Accessed 12 Feb 2016

Energy Charter Secretariat (2011) Follow-up in-depth review of the investment climate and market structure in the energy sector. Energy Charter Secretariat, Brussels, Belgium

Engen OA (2009) The development of the Norwegian petroleum innovation system: a historical overview. In: Fagerberg J, Mowery DC, Nelson RR (eds) Oxford handbook of innovation. Oxford University Press, Oxford, pp 179–207

Farchy J (2015) Azerbaijan devalues currency by one-third amid oil price tumble. The Financial Times, 22 February. Available at: http://www.ft.com/cms/s/0/f7d91efe-ba7f-11e4-945d-00144feab7de.html#axzz3zw63544a. Accessed 12 Feb 2016

Fattahova N (2015) Oil companies affected by prices drop. Kuwait Times, 21 January. Available at: http://news.kuwaittimes.net/oil-companies-affected-prices-drop/

Fearnley TA (2012). Norway's Government Pension Fund: investing for the long run. Paper presented at the Allianz-Oxford pensions conference, Oxford, 27 Sept

FMPR (2015) Federal Ministry of Petroleum Resources. Abuja, Nigeria

Gelb A, Eifert B, Tallroth NB (2002) The political economy of fiscal policy and economic management in oil-exporting countries. Policy research working paper 2899, The World Bank, Washington, DC

George AL, Bennett A (2005) Case studies and theory development in the social sciences. MIT Press, Cambridge, MA

Gojayev V (2010) Resource nationalism trends in Azerbaijan 2004–2009. Working Paper, RUSSCASP, Lysaker, Norway

Gylfason T (2001) Natural resources, education, and economic development. Eur Econ Rev 45(4–6):847–859

Heum P (2008) Local content development: experiences from oil and gas activities in Norway. SNF Working Paper No. 02/08, Institute for Research in Economics and Business Administration, Bergen, Norway

Hou Z, Keane J, Kennan J, Te Velde DW (2015) The oil price shock of 2014. Working Paper, Overseas Development Institute, London, UK

Hovland KM (2014) Norway faces new economic realities with oil's price decline. Wall Street J, 29 October. Available at: http://www.wsj.com/articles/norway-faces-new-economic-realities-with-oils-price-decline-1414607069. Accessed 12 Feb 2016

Humphreys M, Sachs JD, Stiglitz JE (2007) Introduction: what is the problem with natural resource wealth? In: Humphreys M, Sachs JD, Stiglitz JE (eds) Escaping the resource curse. Columbia University Press, New York, pp 1–20

Hunter T (2014) The role of regulatory frameworks and state regulation in optimising the extraction of petroleum resources: a study of Australia and Norway. Extr Ind Soc 1(1):48–58

IEA (2011) Oil and gas security: emergency response of IEA countries. International Energy Agency, Paris, France

ILO (2012) Current and future skills, human resources development and safety training for contractors in the oil and gas industry. International Labour Organization, Geneva, Switzerland

IMF (2012) Kuwait: macroeconomic policy framework for resource-rich developing countries. International Monetary Fund, Washington, DC

IMF (2013) Kuwait: selected issues. IMF Country Report No. 13/337, International Monetary Fund, Washington, DC

IMF (2014a) Kuwait: selected issues. IMF Country Report No. 14/334, International Monetary fund, Washington, DC

IMF (2014b) Nigeria. IMF Country Report No. 14/103, International Monetary Fund, Washington, DC

IMF (2015) Nigeria. IMF country report No. 15/84, International Monetary Fund, Washington, DC

Kjærnet H (2010) The state oil company SOCAR: a microcosm of Azerbaijani development? Caucasus Anal Dig 16:5–7

Kuwait Investment Authority (2015) Investment objectives. Available at: http://www.kia.gov.kw/en/ABOUTKIA/Pages/OBJECTIVES-PRINCIPLES.aspx

Lahn G, Marcel V, Mitchell J, Myers K, Stevens P (2009) Good Governance of the National Petroleum Sector: the Chatham House document. The Royal Institute of International Affairs, London

Leskinen O, Bekken PK, Razafinjatovo H, García M (2012) Norway oil and gas cluster: a story of achieving success through supplier development. Harvard Business School, Cambridge, MA. Available at: http://www.isc.hbs.edu/resources/courses/moc-course-at+-harvard/Documents/pdf/student-projects/120503%20MOC%20Norway%20final.pdf. Accessed 12 Feb 2016

Lijphart A (1971) Comparative politics and the comparative method. Am Polit Sci Rev 65 (03):682–693

Marcel V (2015) Guidelines for good governance in emerging oil and gas producers. Chatham House, London

Maxwell JA (2004) Causal explanation, qualitative research, and scientific inquiry in education. Educ Res 33(2):3–11

McGroarty P, Hinshaw D, Akingule G (2015) Nigeria may double VAT tax because of oil's drop. Wall Street J, 20 January. Available at: http://www.wsj.com/articles/nigeria-keeps-key-interest-rate-on-hold-1421768219. Accessed 12 Feb 2016

NNPC (2015) Nigeria. Nigerian National Petroleum Corporation, Abuja, Nigeria

NPD (2015) Facts 2013: the Norwegian Petroleum Sector. Norwegian Petroleum Directorate, Stavanger, Norway

NSIA (2015) Nigeria Sovereign Investment Authority. Available at: http://nsia.com.ng/overview/. Accessed 12 Feb 2016

OECD (2007) Economic survey of Norway 2007: the petroleum sector and its impact. Organization for Economic Co-operation and Development, Paris, France

OECD (2014) OECD skills strategy diagnostic report: Norway. Organization for Economic Co-operation and Development, Paris, France

Omojimite BU (2011) Building human capital for sustainable economic development in Nigeria. J Sustain Dev 4(4):183

OPEC (2014) Nigeria facts and figures. Organization of the Petroleum Exporting Countries, Vienna, Austria

Ovesen V (2008) Petroleum fiscal regimes and revenue management systems. Paper presented at African Development Bank seminar, Kampala, Uganda, 8–9 July

PTDF (2015) Overseas scholarship scheme, Nigeria. Petroleum Technology Development Fund. Available at: http://www.ptdf.gov.ng/scholarship. Accessed 12 Feb 2016

Ragin CC (1987) The comparative method: moving beyond qualitative and quantitative strategies. University of California Press, Berkeley, CA

Reuters (2015) Kuwait slashes spending plan, sees big deficit in 2015/16 budget. Gulf business, 27 January. Available at: http://www.gulfbusiness.com/articles/industry/kuwait-slashes-spending-plan-sees-big-deficit-in-2015/16-budget/. Accessed 12 Feb 2016

Ross ML (1999) The political economy of the resource curse. World Polit 51(2):297–322

Rosser A (2006) Escaping the resource curse. New Polit Econ 11(4):557–570

Scott J (1990) A matter of record: documentary sources in social research. Polity Press, Cambridge

El-Katiri L, Fattouh B, Segal, P (2011) Anatomy of an oil-based welfare state: rent distribution in Kuwait. Research Paper 3, Kuwait programme on Development, Governance and Globalisation in the Gulf States, London School of Economics, London

Snyder R (2001) Scaling down: the subnational comparative method. Comp Int Dev 36(1):93–110

SOFAR (2014) The stat oil of the Azerbaijan Republic. Available at: http://www.ifswf.org/members/azerbaijan. Accessed 12 Feb 2014

Stake RE (1995) The art of case study research. Sage, Thousand Oaks, CA

Statoil (2015) Major shareholders, Annual report 2011. Statoil, Stavanger, Norway. Available at: http://www.statoil.com/annualreport2011/en/shareholderinformation/pages/majorshareholders.aspx. Accessed 12 Feb 2016

Stevens P (2008) Kuwait Petroleum Corporation: searching for strategy in a fragmented oil sector. Working Paper 78, Program on energy and sustainable development, Stanford University, Stanford, CA

Stevens P, Dietsche E (2008) Resource curse: an analysis of causes, experiences and possible ways forward. Energy Policy 36:56–58

Stevens P, Lahn G, Kooroshy J (2015) The resource curse revisited. Research Paper, Chatham House, Royal Institute oF International Affairs, London

SWFI (2015a) Sovereign Wealth Fund, Kuwait. Available at: http://www.swfinstitute.org. Accessed 12 Feb 2016

SWFI (2015b) Sovereign Wealth Fund, Norway. Available at: http://www.swfinstitute.org. Accessed 12 Feb 2016

SWFI (2015c) State Oil Fund of the Azerbaijan Republic, Azerbaijan. Available at: http://www.swfinstitute.org. Accessed 12 Feb 2016

Thurber MC,Istad BT (2010) Norway's evolving champion: Statoil and the politics of state enterprise. Working Paper 92, Program on Energy and Sustainable Development, Stanford University, Stanford, CA

Thurber MC, Emelife IM, Heller PR (2010a) NNPC and Nigeria's oil patronage ecosystem. Working Paper 95, Program on energy and sustainable development, Stanford University, Stanford, CA

Thurber M, Hults D, Heller PR (2010b) The limits of institutional design in oil sector governance: exporting the 'Norwegian Model'. Paper presented at the International Studies Association Annual Convention, New Orleans, LA, 18 February

Thurber MC, Hults DR, Heller PR (2011) Exporting the 'Norwegian model': the effect of administrative design on oil sector performance. Energy Policy 39(9):5366–5378

Tordo S, Warner M, Manzano O, Anouti Y (2013) Local content policies in the oil and gas sector. World Bank Publications, Washington, DC

Van der Ploeg F, Poelhekke S (2009) Volatility and the natural resource curse. Oxf Econ Pap 61 (4):727–760

Victor DG, Hults DR, Thurber MC (eds) (2011) Oil and governance: state-owned enterprises and the World energy supply. Cambridge University Press, Cambridge

Wallis W (2014) Nigeria warns on falling oil prices. Financial Times, 26 October. Available at: http://www.ft.com/cms/s/3aad3da4-5ba9-11e4-a674-00144feab7de. Accessed 12 Feb 2016

Watts M (2004) Resource curse governmentality, oil and power in the Niger Delta. Nigeria Geopolitics 9(1):50–80

Wolf CO, Pollitt MG (2009) The welfare implications of oil privatization: a cost-benefit analysis of Norway's Statoil

World Bank (2007) Investing in oil in the Middle East and North Africa: institutions, incentives and the national oil companies. Sustainable development department paper 40405-MNA, World Bank, Washington, DC

World Bank (2015a) Azerbaijan partnership program snapshot. World Bank Group, Washington, DC. Available at: http://pubdocs.worldbank.org/pubdocs/publicdoc/2015/10/409231443709836813/Azerbaijan-Snapshot.pdf. Accessed 12 February 2016

World Bank (2015b) World development indicators. Available at: http://data.worldbank.org/indicator. Accessed 12 Feb 2016

Yin RK (2003) Case study research: design and methods, 3rd edn. Sage Publications, Thousand Oaks, CA

Chapter 5
Importance and Development of the Oil and Gas Industry in the Iraqi Kurdistan Region

Abstract The Kurdistan region has plentiful oil and gas reserves. Despite the opportunities generated by the oil and gas industries, they may also bring risks of high economic dependence on the oil and gas sector due to KRG's failure to diversify the economy away from resource extraction towards productive and sustainable activity, as well as a worsening of the relationship between the people and the government, through economic distortion, lack of transparency and accountability, increased environmental pollution and internal tensions. Main findings show that oil and gas discovery and production lead the people of the Kurdistan region to expect jobs and prosperity. To the extent that these are not realised, the risk of social unrest may increase.

Keywords Kurdistan region's petroleum sector · Transparency · Accountability · Oil revenue management · Economic diversification

5.1 Introduction

In recent years, the oil and gas sector in the Kurdistan region has received significant political and strategic attention, not only because of the implications of increased oil and gas activities for the relationship between the Kurdistan region and the Iraqi central government (The Guardian 2014; The Economist 2012, 2014), but also because of the sector's huge contribution to the economy of the region (Oil and Gas Year 2014; Ministry of Natural Resources 2015a).

As in the rest of Iraq, the Kurdistan region lacks infrastructure, production capacity and capability in agriculture and industry, which may be the main reason for its continuing high dependence on the extraction of natural resources. The oil and gas industry is the driving force behind development in many areas in the region, such as electricity, water supply, healthcare, education, telecommunications and housing.

K. Abdullah Auzer, *Institutional Design and Capacity to Enhance Effective Governance of Oil and Gas Wealth: The Case of Kurdistan Region*, Perspectives on Development in the Middle East and North Africa (MENA) Region, DOI 10.1007/978-981-10-4518-9_5

A qualitative case study approach is used in this chapter to answer the third research question of the book, which examines the pattern of development of oil and gas by the Kurdistan Regional Government (KRG) and its role in the wider economy. The Kurdistan region offers an interesting opportunity to study oil and gas policy, as it has chosen a different path from its central government counterpart in Baghdad in terms of management of the industry. In this study, the term 'oil and gas policy' refers to the framework of regulations, fiscal devices, institutional arrangements and procedures that constitute the environment of Kurdistan's oil and gas industry, which explores, develops, produces, transports and markets oil and gas products. Since the oil industry and its management style are in the early stages of development in the Kurdistan region, it is of particular interest to observe how it is evolving.

Oil and gas is a relatively new sector in the Kurdistan region and, as such, provides a unique opportunity to assess how policy instruments (with respect to the fiscal regime and human resources) and administrative arrangements interact to influence the pace of development. In addition to highlighting both shortcomings and positive aspects of this newly-established industry, this study may be used as a potential reference for studying other regions and states in similar circumstances.

A within-case analytic strategy has been used, based on thematic analysis (Yin 2009), to explore the development of this new petroleum sector and its contribution to the growth of other economic sectors. Three themes, based on the literature review and findings presented in Chaps. 3 and 4, delimit the scope of this study: the development of the economy, the unemployment rate, and the oil and gas industry in the Kurdistan region.

Secondary data sources are used in this chapter. As explained in Chap. 4, general rules have been considered with respect to evaluating the dataset. Data and statistics published by the ministries of Planning and Natural Resources and other related ministries in the Kurdistan region, together with published surveys in the public domain, have been used to assess the development of the oil and gas sector and the wider economy of the Kurdistan region.

This chapter is organised as follows. Section 5.2 charts KRG's economic growth and the role of public and private investment. Section 5.3 examines factors affecting the unemployment rate in the region. Section 5.4 focuses on the development of KRG's oil and gas sector, from exploration to production and distribution, infrastructural enlargement and the more complex process of institutionalisation of the oil and gas industry. The final section summarises the findings of the chapter, setting the context for the next chapter, which will present an in-depth analysis of the opportunities and challenges facing the region's petroleum industry and wider economy.

5.2 Economic Development in the Kurdistan Region

The whole of Iraq is heavily dependent on its petroleum revenues: the proceeds of the oil and natural gas industry account for over 90% of the federal budget (IEA 2012). Iraqi public income has been dramatically affected by the oil price collapse in 2014 and by ongoing conflict within the country (Coles et al. 2015), and uncertainty over oil revenues has also had an impact on KRG's budget.

In principle, the Kurdistan region's budget is allocated annually by the central government once the Iraqi parliament has approved the national budget. The region's budget is about 17% of the federal budget, reflecting the percentage share of Kurdistan's population within the state of Iraq. This fiscal arrangement was laid out in the 2005 constitutional amendment (UNDP 2009); however, the KRG has constantly complained that the allocated budget has never been received in full (UNDP 2009). Uncertainties in both the timing and the transfer of the actual amounts of its budget entitlement have significantly affected economic development in Kurdistan. This dependence on fiscal transfers from the federal government makes the Kurdistan region subject to political pressure from Baghdad, which has used the region's budget share as a basis for negotiation in ongoing verbal disputes between the federal government and KRG (Al-Salhy and Lyon 2014; Passche and Mansurbeg 2014).

Budgetary transfers from Baghdad to Kurdistan constitute a major source of the region's income and contribute directly to its economic growth. As a result of an improved security situation and a more business-friendly environment than in the rest of Iraq, the Kurdistan region experienced remarkable economic growth between 2003 and 2011. The region's GDP increased from US$0.6 billion in 2003 to US$23 billion in 2011 (see Fig. 5.1).

However, no reliable statistics are available to inform research on the level and growth rate of GDP over the period covered by this study. The only accessible

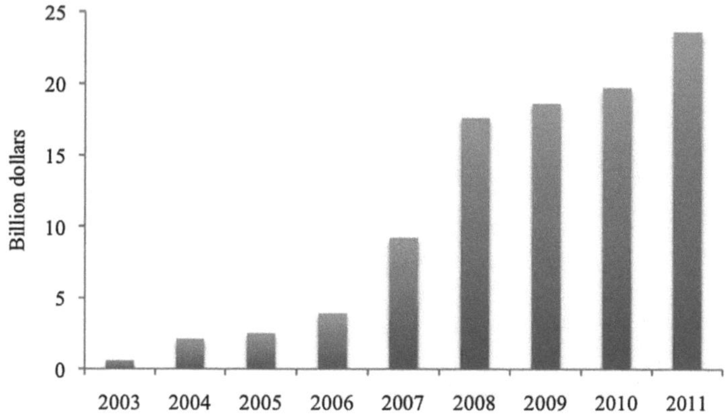

Fig. 5.1 Development of the Kurdistan region's GDP, 2003–2011. *Source* KRSO (2015)

statistics are from a report published by the Kurdistan Region Statistics Office (KRSO) with respect to the contribution of different economic sectors to overall GDP in 2011.

Productive sectors such as agriculture and manufacturing have not grown in Kurdistan since 2011. Examination of the relative shares of different economic sectors in the Kurdistan region reveals that the service sector (public administration and other services) forms 50% of GDP, while the contributions of productive sectors such as manufacturing and agriculture are 9.4 and 17.5% respectively (see Fig. 5.2). It can be concluded that the manufacturing and agricultural sectors, as determinants of economic diversification in the Kurdistan region (USAID 2008; Heshmati 2012), show much lower growth rates than the service sector. Furthermore, KRG's expenditure on and investment in infrastructure are critical components of GDP growth, and play a major role in increasing outputs across the various sectors.

5.2.1 Public Capital Investment

The government's total capital expenditure is a critical indicator of the contribution of public investment to the Kurdistan regional economy, and is funded from two sources. The first is KRG's revenues from the federal budget and non-oil revenues generated within the region, such as administration tax, import/export taxes, income tax and customs/duties. According to the Economic and Finance Committee (2013), fiscal transfers from the federal budget made up 96% of KRG's revenues in the 2013 budget. The second source is Kurdish oil revenues, which are transferred into a separate account outside the region's budget and are managed by the Ministry of Natural Resources (MNR) in Kurdistan (Ministry of Natural Resources 2014).

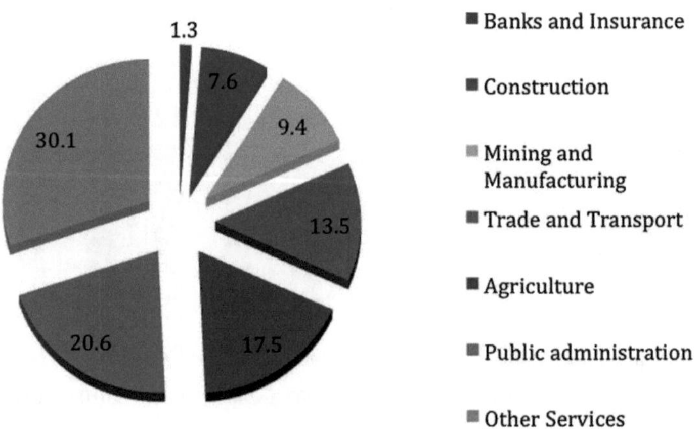

Fig. 5.2 Distribution of annual growth rates in valued added of non-oil sectors of the Kurdistan region's economy, 2011. *Source* KRSO (2015)

Relatively little budgetary income has been channelled towards investment projects. The government's non-capital and capital expenditure accounted for nearly 70 and 30% respectively of the total budget for the period 2010–2013 (Economic and Finance Committee 2012, 2011, 2013; KRSO 2015). This indicates an imbalance between revenues and expenditures, or between producing and consuming sectors. On the expenditure side, government spending falls into three main categories: public servants' wages, public recurrent non-wages (non-capital spending) and capital expenditure.

However, public capital expenditure increased during the period 2009–2013 (see Fig. 5.3), from US$89 million in 2007 to about US$4 billion in 2013. Figure 5.4 also highlights that in 2014 investment in Kurdistan was dramatically affected by the Iraqi central government freezing the region's share of the national budget. Political disputes between Baghdad and Erbil stem from continuing problems regarding the management of Kurdistan's oil and gas industry (Holland 2012), as a result of which KRG has been unable to decrease its dependence on fiscal transfers (oil revenues) from the central government.

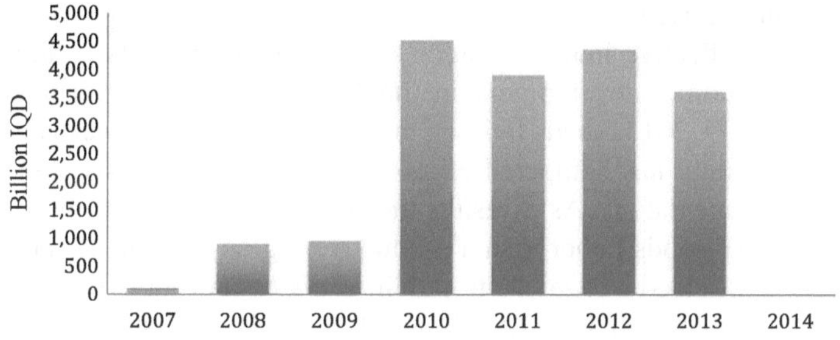

Fig. 5.3 Kurdistan region's public capital investment, 2007–2013. *Source* Ministry of Planning (2015)

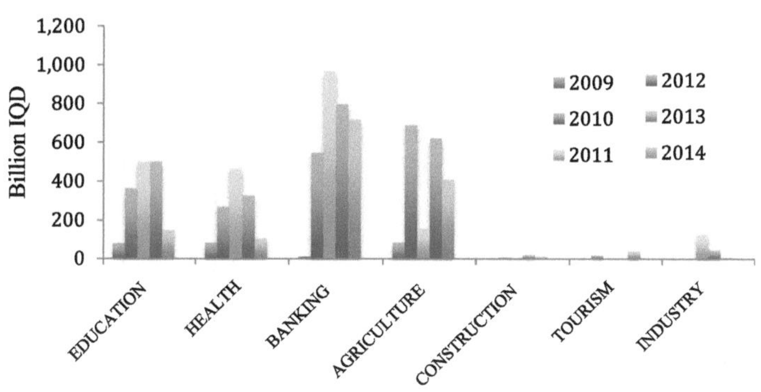

Fig. 5.4 Kurdistan region's public capital investment across different sectors. *Source* Ministry of Planning (2015)

As illustrated in Fig. 5.4, the level of government funding for agriculture has been relatively high compared to other sectors in recent years; however, the budgetary allocation to this sector has been inconsistent. The Kurdistan region has suffered from under-investment in industry, resulting in a very low contribution to GDP of the productive manufacturing sector, including the agricultural products processing industry (see Fig. 5.4).

In addition to budgetary income, Kurdish oil revenues (see Fig. 5.5) have also been allocated by MNR to support various economic sectors (Ministry of Natural Resources 2013a, b). In contrast to regional budget allocations, a limited amount of capital expenditure from oil revenues has been invested to develop education and support the agricultural sectors. A large proportion of oil revenues have been spent on purchasing fuel and fuel products to meet local consumption needs (see Fig. 5.6).

This section has examined trends in public investment expenditure in different sectors of the Kurdistan region's economy. The results show that public expenditure on agricultural and industrial investment projects, as a proportion of total public investment expenditure is low. This is inconsistent with KRG's agricultural and industrial policy, as elaborated in its medium-term development plan (2012–2016), which prioritises the development of the agricultural and industrial sectors (Heshmati 2012).

In addition, ineffective implementation of investment projects is another major hindrance to economic development in the Kurdistan region. A United Nations report (UNDP 2009, 2013) sheds light on the fact that a lack of checks and balances due to poor oversight has hampered efforts, resulting in ineffective and inefficient investment implementation. As a result, the Kurdistan region has become a consumer of imported goods rather than a productive region capable of catering for its own basic needs, added to which its public services are poor. For example, the electricity industry, which is under the control of the ministry of electricity, is heavily dependent on the government financial support as a result of very low

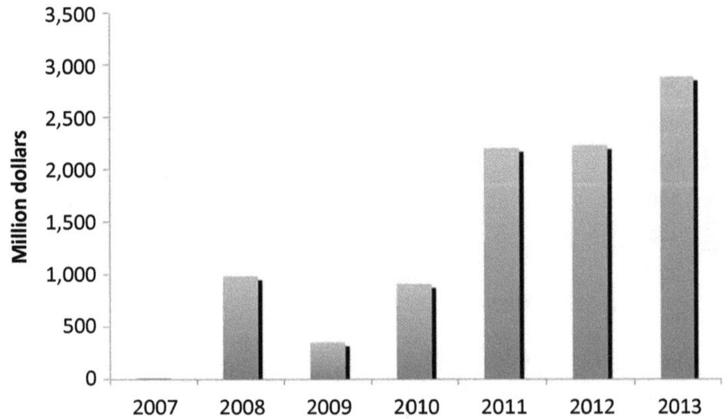

Fig. 5.5 Kurdistan region's oil revenues, 2007–2013. *Source* Ministry of Natural Resources (2013a, b)

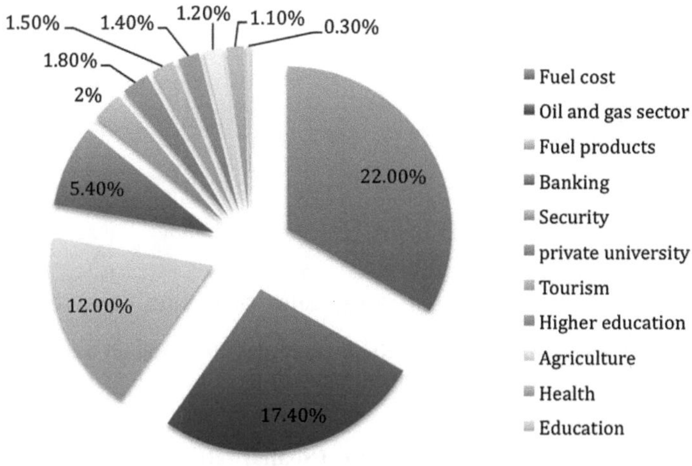

Fig. 5.6 Distribution of Kurdistan region's oil revenue expenditure, 2013. *Source* Ministry of Natural Resources (2013a, b)

electricity tariff and high electricity losses. The ministry of electricity's budgetary support in form of financial subsidies (the difference between the direct cost faced by the government ministry) and operational subsidies (the differences tariffs and economic costs of fuels) are annually about US$3 billion annually. This puts heavy pressure on the Kurdistan region's annual budget (UNDP 2012; Ekurd Daily 2015) but, despite such an outlay, efficient and adequate supplies of electricity remain a major challenge in the region (Ekurd Daily 2015).

5.2.2 *Private Capital Investment*

The KRG claims that one of its aims is to increase foreign investment in Kurdistan. An investment law enacted in 2006 to create a business-friendly environment for foreign companies in the Kurdistan region (Kurdistan Regional Government 2015) stipulates that foreign investors are permitted to repatriate their profits freely, are treated equally with national investors under the law, are entitled to all the capital assets of any project, are exempt from taxes, duties and the requirement to obtain import licences, and enjoy the same rights as local investors to purchase and own land. It also establishes additional tax incentives and benefits to encourage foreign entrepreneurs. However, this law covers all economic sectors except the oil and gas sector. The priority sectors appear to be agriculture, industry and tourism (Kurdistan Board of Investment 2015).

Since 2014, private investment has declined. Total private investment fell by about 66% in 2014 compared with the previous year (see Fig. 5.7). Again, this can be attributed to the fiscal crisis caused by the federal government freezing all

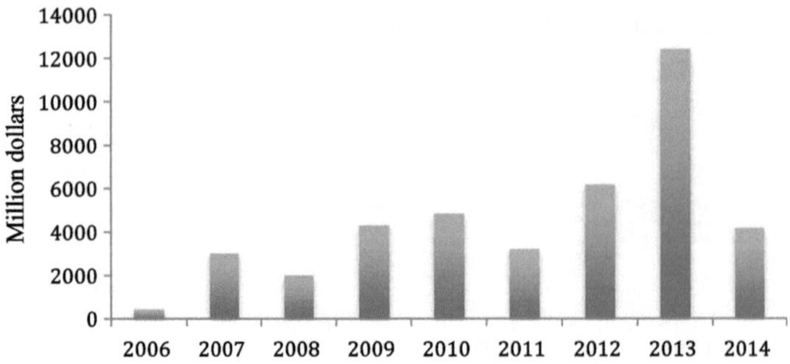

Fig. 5.7 Private capital investment in the Kurdistan region (excluding oil and gas investments), 2006–2014. *Source* Kurdistan Board of Investment (2015)

payments to KRG after January 2014 (Shafaq 2014; Rudaw 2014). Private investment also fell owing to the war against the so-called Islamic State of Iraq and Syria, also known as ISIS (World Bank 2015a). These two factors have negatively affected the flow of private investment into the Kurdistan region.

In common with public capital investment priorities, private investment in productive sectors such as manufacturing and agriculture is also very low compared with the construction sector. Private sector investment falls into three types in the Kurdistan region— national, joint venture and foreign investment, accounting for 13.75, 9.07 and 77.18% of total private investment respectively. Since 2006, private funding has been directed primarily towards housing, tourism, industry and trade, with the importation of durable and non-durable goods clearly dominating (see Fig. 5.8). The low contribution of private investment to capacity building and diversification of the economy indicates that KRG has failed to encourage domestic and foreign investors to engage in financing non-oil productive sectors in Kurdistan.

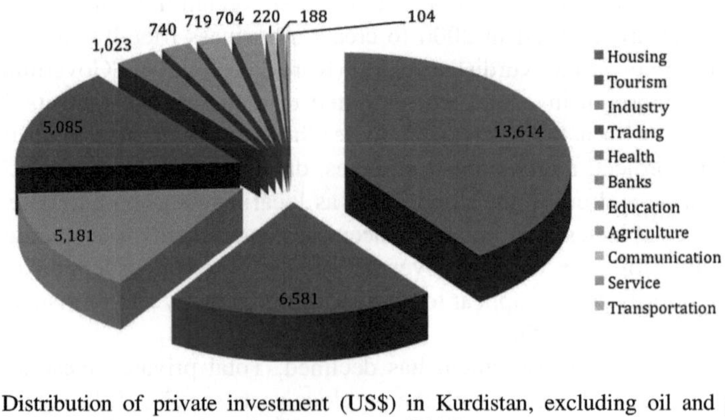

Fig. 5.8 Distribution of private investment (US$) in Kurdistan, excluding oil and gas sector, 2006–2014. *Source* Kurdistan Board of Investment (2015)

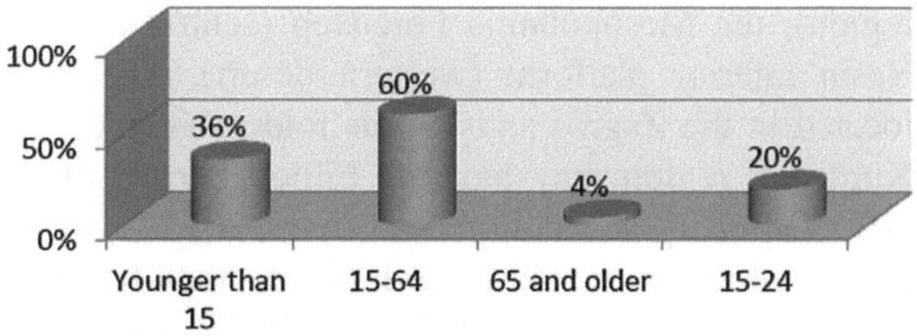

Fig. 5.9 Kurdistan region's population. *Source* KRSO (2014)

5.2.3 Unemployment Rate in the Kurdistan Region

The Kurdistan region's population reached 5.2 million in 2014. It is a young society, with 36% aged 0–14 years, 20% aged 15–24, and only four per cent aged over 63 years (Fig. 5.9). In recent years, the percentage of potentially economically active people has increased for both males and females from 38.4% in 2012 to 39% in 2013 (KRSO 2014).

In 2012, the unemployment rate in the Kurdistan region was 7.9%, which is lower than the overall level of 8.6% in Iraq (KRSO 2014). The public sector currently employs the majority of the working population in Kurdistan. Employment rates in the public and private sectors are around 50 and 51% respectively. These are aggregate rates covering various economic sectors. The service sector registers the highest employment rate (77.3%), followed by industrial activities (16.6%), while the lowest rate (6.1%) is for agricultural activities (KRSO 2012).

Given this context, public wage payments continue to exert significant pressure on the regional government's budget, resulting in the crowding out of important public investment programmes. The major obstacle hindering the creation of non-public sector jobs and opportunities for the economically active population is a shortage of necessary skills and expertise (Heshmati 2007; UNDP 2012). The KRG's human resource development policy has failed to establish "a direct link between a successful investment strategy and a successful skills supply system that is responsive to market needs for skills in the labour force" (UNDP 2012, p. 92). The main reasons for this are the poor quality of higher education, the absence of a vocational training system and a mismatch between the supply of suitably educated people and the skills demanded by the labour market. An uncompetitive workforce has led to an increase in the unemployment rate among young people in the Kurdistan region (Heshmati 2007, 2012; UNDP 2012).

5.3 Development of the Oil and Gas Industry

The Kurdistan region has plentiful oil and gas reserves. According to British Petroleum (2015), Iraq's proven oil reserves are estimated to be between 141 and 150 billion barrels, making it one of the world's top holders of proven oil reserves. Iraq's main hydrocarbon basins are located in the Zagros fold belt (Kurdistan region), the Mesopotamia Foredeep (central and southern Iraq), and the Widyan Basin interior platform (western desert). The Kurdistan region's resources are located in the Zagros and Taurus fold belts (see Fig. 5.10). It is estimated that the Kurdistan region may have 45 billion barrels of oil reserves (Oil and Gas Year 2014; Ministry of Natural Resources 2015a), which, if Kurdistan were an independent country, would place it among the 10 richest countries in terms of oil and gas reserves (Pfluger and Duero 2011). This does not include reserves in the super-giant field near Kirkuk, an area disputed by the Kurdish region and the central government in Baghdad (Gray 2012).

In short, the Kurdistan region of Iraq is one of the world's largest remaining onshore oil frontiers. The region has gone through several phases of exploration since the first well was drilled in 1901 in the Chia Surkh area, which was also the first well ever, drilled in the Middle East (Mackertich and Samarrai 2015). After the fall of Saddam Hussain, and with further strengthening of KRG against the central government, the region witnessed a dramatic boom in exploration activities, and many international oil companies entered the region (Oil and Gas Year 2014; Ministry of Natural Resources 2015a).

In recent years, well-known companies such as Exxon Mobil (US) and Total France have entered the exploration race. At the same time, smaller firms have also achieved some significant breakthroughs, not only by discovering and producing Kurdish oil, but also by actually managing to deliver it to the international market (Ministry of Natural Resources 2015a). Since 2005, over 180 wells have been drilled in the region and, despite fluctuating oil prices, exploration activities are still

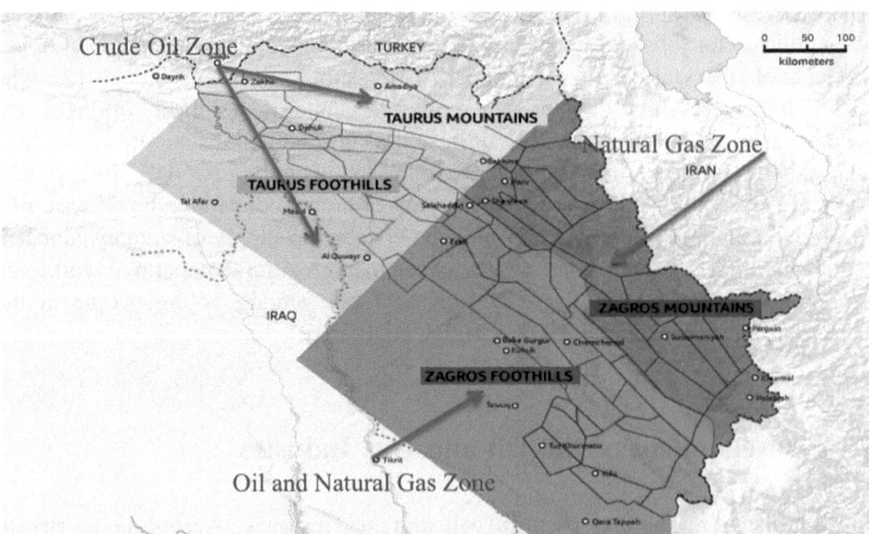

Fig. 5.10 Structural domain of Kurdistan region's oil and gas resources. *Source* Ministry of Natural Resources (2015a)

thriving. The region has always been considered risky, not only because of political uncertainty, but also because of its different geological setting from the rest of Iraq (Mackertich and Samarria 2015). However, the success of the recently discovered fields in the Kurdistan region has opened new horizons for de-risking deeper targets previously considered non-productive, both in Iraq as a whole and in the Kurdistan region (Mackertich and Samarria 2015).

Direct foreign investment by multinational energy companies is the main driver of oil and gas industry development in the region (Oil and Gas Year 2014; Ministry of Natural Resources 2015a). One perceived advantage of making deals with KRG is the fact that the region has its own petroleum law, which was approved by the Kurdistan region's parliament in 2007. This has stimulated new foreign private sector participation in Kurdistan's oil and gas sector (Ministry of Natural Resources 2015a). Contracts signed between KRG and foreign companies are based on a product-sharing contract (PSC) model (Ministry of Natural Resources 2015a). By the end of 2014, 50 PSCs had been issued by KRG, as presented in Appendix 1 (Oil and Gas Year 2014; Ministry of Natural Resources 2015a).

5.3.1 Oil Production

As can be seen from Fig. 5.11, since 2007, KRG has successfully increased oil production capacity (Ministry of Natural Resources 2015a). There are three key oil fields in Kurdistan capable of significant production: Tawke (operated by DNO), Taq Taq (operated by Genel Energy and Addax Petroleum) and Shaikan (operated by Gulfkeystone). Over the period 2003–2015, oil production has grown steadily and sustainably. In December 2015, the crude oil production capacity of fields operating in the KRG was 637,137 monthly average barrels per day.

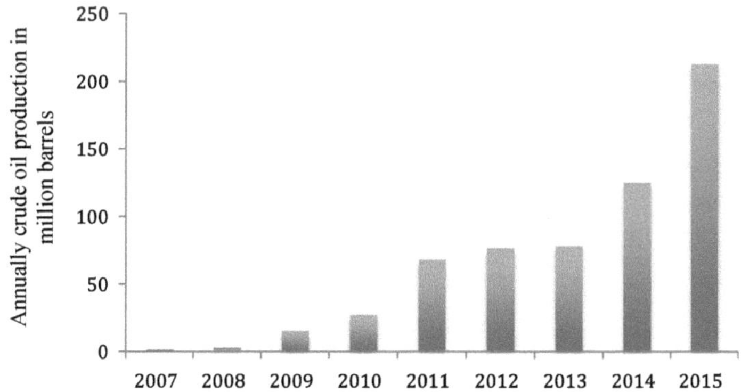

Fig. 5.11 Kurdistan region's oil production, 2003–2015. *Source* Based on data from Ministry of Natural Resources (2014) and (2015b)

The Kurdistan region's share of total domestic oil consumption is 17% of Iraq's total oil domestic consumption. However, since 2004, the KRG has faced a historical and systematic shortfall in its entitlement of domestic oil consumption. Therefore, the KRG continues a policy of supplying the needed refined products such as diesel, benzene and kerosene to domestic demand by swapping and selling crude oil, as illustrated in Fig. 5.12.

5.3.2 Oil Exports

Kurdish oil has found its way to the global market. Kurdistan has been exporting its oil through a pipeline between Kirkuk and the Turkish port of Ceyhan, but this continues to be controlled by Baghdad. The federal government's marketing organisation, SOMO, was directly responsible for controlling it from 2008 to 2012 (see Fig. 5.13).

The KRG has a goal of marketing its crude oil independently and, to this end, has finished building its own pipeline to Turkey, the Taq Taq-Khurmala-Fish Khabur pipeline (see Fig. 5.14). This new pipeline has significantly boosted the Kurdistan region's oil exports and has also strengthened Erbil's role by linking it to the Kirkuk oilfield (Burr2014; Barbosa 2014).

Official KRG crude exports have faced serious challenges. The federal government's unwillingness to further KRG's independence in the development of its oil and gas industry has resulted in political tension between Erbil and Baghdad. As a result, Kurdistan's budget, as a constitutional entitlement, has been cut by the central government (Rudaw 2014), increasing political tension still further.

In December 2014, the Kurdistan region signed a new oil agreement with Iraq's federal government, aiming to put an end to years of conflict over oil and gas

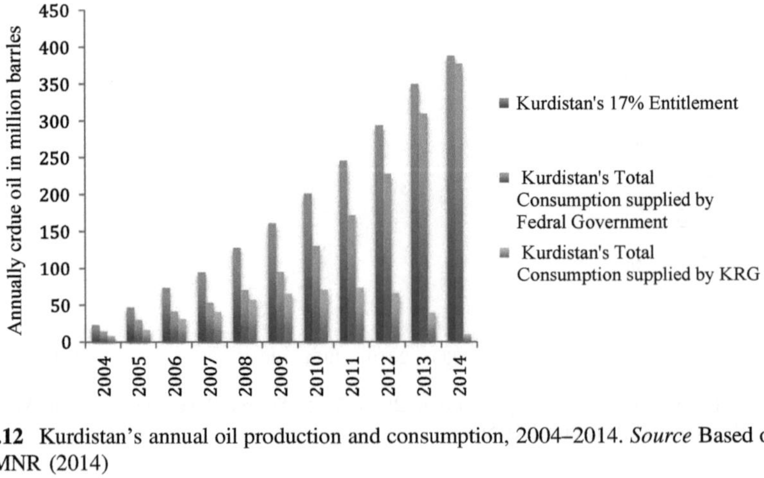

Fig. 5.12 Kurdistan's annual oil production and consumption, 2004–2014. *Source* Based on data from MNR (2014)

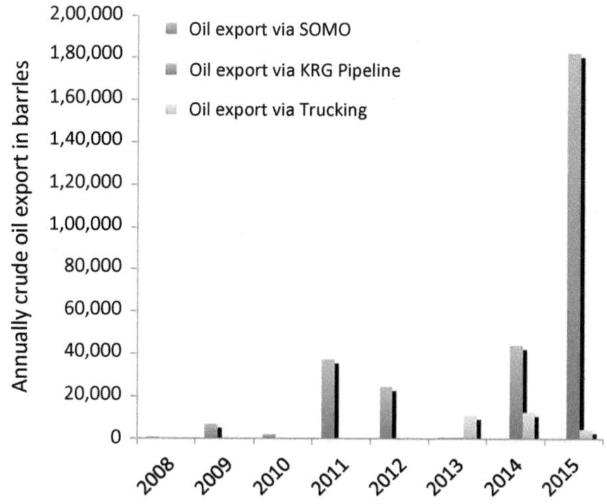

Fig. 5.13 Kurdish oil exports via trucking, KRG pipeline and SOMO. *Source* Based on data from Ministry of Natural Resources (2014) and (2015b)

management. Under this oil deal, KRG will supply 550,000 bpd of oil for sale under Baghdad's supervision. This deal includes the export of 300,000 bpd from Kirkuk oil fields, which were added to the producing fields in the Kurdistan region after Kurdish forces regained the city of Kirkuk from ISIS (Hawramy and Beaumont 2014). In addition to Kirkuk's oil, KRG also sends 250,000 bpd of Kurdish oil through the KRG-Turkey pipelines. In return, KRG receives its 17% share of Iraq's total budget (Coles et al. 2015).

5.3.3 *Refineries*

Kurdistan's downstream sector has been developing through the involvement of the private investment (Ministry of Natural Resources 2015a). The Kurdistan region has two officially recognised refineries. The Kalak refinery, operated by the KAR Group, is the largest oil refinery in the Erbil governorate, and currently has a capacity of 100,000 bpd. Efforts are ongoing, with the help of international expertise, to expand this capacity to 200,000 bpd in the near future. Secondly, the Qaiwan Group operates the Bazian refinery, located in the Sulaimaniya area, with a capacity half that of the Kalak refinery. The Qiwan Group intends to double its refined product capacity in the coming years (Ministry of Natural Resources 2015a). Despite the initiatives mentioned above, the fuel crisis remains unresolved in Kurdistan. The refinery sector faces constant challenges, including the severe shortage and poor quality of refined products such as gasoline. As a result, both the public and Kurdistan's members of parliament have frequently criticised the quality of the gasoline produced, questioning why there are no major public refineries that could deliver products in accordance with internationally-recognised standards (Ministry of Natural Resources 2015a).

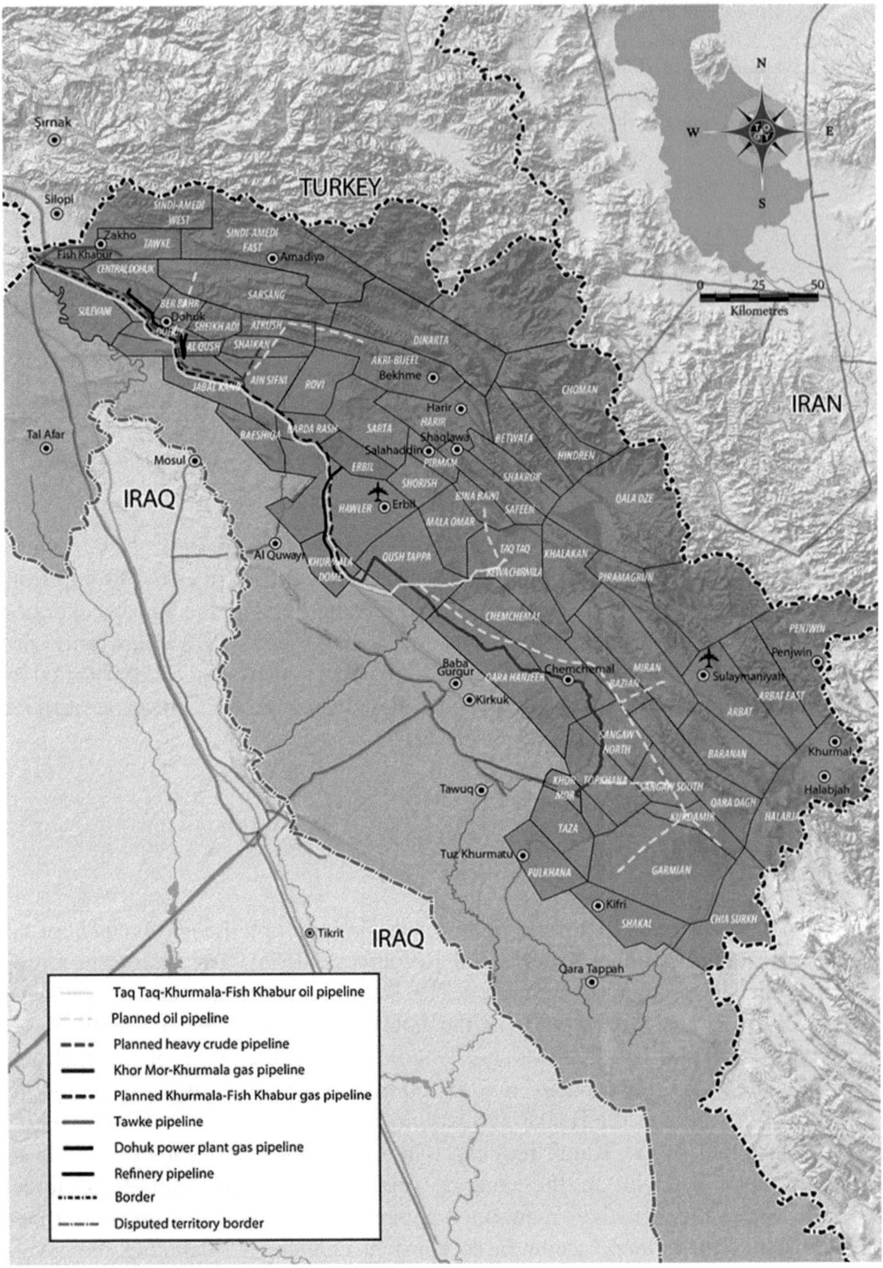

Fig. 5.14 Kurdistan region's pipeline infrastructure. *Source* Ministry of Natural Resources (2015a)

5.3.4 Natural Gas

The Kurdistan region's giant natural gas field has made the region a potential key player in global energy security (Dickel et al. 2014; Gray 2012; Chyong et al. 2015). It is estimated that the region holds approximately 708 billion cubic metres of proven gas reserves, and an estimated 2.8–5.6 trillion cubic metres of unproven gas resources (Oil and Gas Year 2014; Ministry of Natural Resources 2015a). The Bina Bawi and Miran gas fields are the most significant in the Kurdistan region, holding an estimated 238 billion cubic metres of natural gas reserves between them (Invest In Group 2014). Genel Energy was awarded a production-sharing agreement by KRG for these fields. In addition, Khor Mor and Chamachamal fields produce an average of 9.5 million cubic metres per day. Natural gas produced by Crescent Petroleum and its affiliate Dana Gas of the United Arab Emirates, OMV of Austria and MOL of Hungary supply to power stations in the Sulymani and Erbil areas (Crescent Petroleum 2015). DNO and Genel Energy also deliver natural gas from the Summail field to a power station in the Dohouk area (Ministry of Natural Resources 2015a). The replacement of diesel-fired power plants with plants fuelled by natural gas is expected to provide KRG with savings on expensive fuel costs, as well as bring environmental benefits.

In addition to domestic demand, natural gas is an alternative supply source for Turkey and Europe. The Kurdistan region's natural gas production could be a possible future source of diversification of EU's natural gas supply through the Southern Gas Corridor, which aims to decrease European dependence on Russian gas (see Fig. 5.15). The Southern Gas Corridor is a project designed to transport natural gas from Caspian and Middle East region to Europe across Turkish territory. The project includes the Southern Caucasus pipeline, the Trans-Anatolian Natural Gas Pipeline (TANAP), and the Trans-Adriatic Pipeline (TAP) (Finon 2011; Belkin et al. 2013; Winrow 2013). Chyong et al. (2015) has identified the need for diversification of the EU's natural gas supply. The current conflict in Ukraine has triggered a need for the EU to seek alternative natural gas supplies in order to reduce its dependence on Russia (Gray 2012), and the Kurdistan region is now at the top of the list of potential supply sources (Chyong et al. 2015).

5.3.5 Fiscal Regime of Kurdistan's Oil and Gas Sector

The current fiscal regime of the region's petroleum sector is based on PSCs. The Kurdistan oil and gas law allows KRG to establish PSCs for the exploration, development and production of oil and gas, while the Federal Government of Iraq favours a risk service contract (RSC) regime (Ministry of Natural Resources 2015a). Under the PSC model, KRG, as the owner of the petroleum, engages international oil companies (IOCs) as contractors to provide technical and financial services for exploration and development operations. The IOCs carry the entire

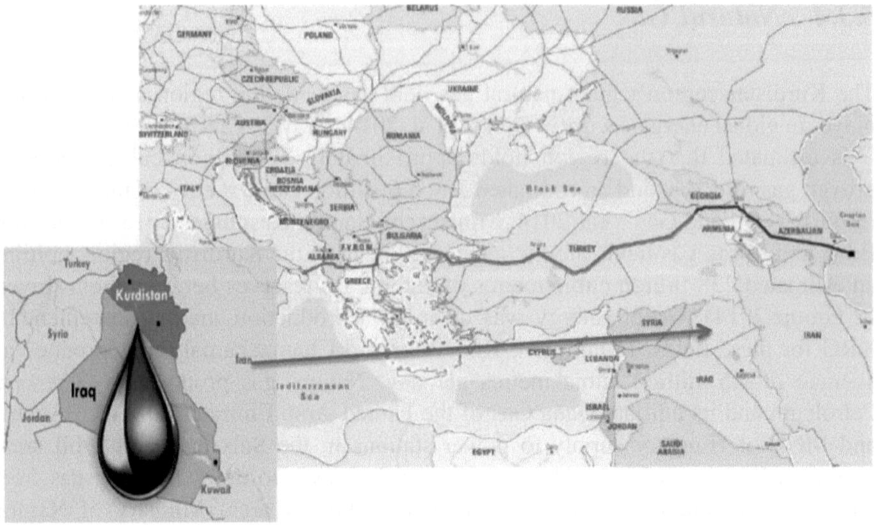

Fig. 5.15 Southern gas corridor pipeline. *Source* The Jamestown Foundation (2013)

exploration risk. If the companies find no commercial hydrocarbon accumulations, they are not entitled to any sort of compensation (Ministry of Natural Resources 2015a). When petroleum is produced, the produced oil and gas is shared between the IOC and the host country. The basic elements of a PSC are royalty oil, cost recovery oil and profit (Sunley et al. 2003). The IOCs pay a royalty of 10% of gross total production to KRG. Cost recovery is determined as a proportion of final production after deduction of the royalty, not exceeding 40% for crude oil and 60% for natural gas (Ministry of Natural Resources 2015a). After deducting the royalty and cost recovery, the remainder of the proceeds is profit.

In general, the share of profit oil produced is based on a factor ‘R’, which is the ratio of cumulative oil revenues earned to cumulative costs incurred by the contractor. The share of profit oil between KRG and the IOC is based on a function of a stepwise increasing value of the R factor as follows. R is less than one when a petroleum contract first begins; in other words, the IOC is incurring costs but is not yet producing. R reaches one once project revenues equal project costs, and once R is larger than one, then the project revenues are greater than project costs, which means the petroleum contract is in the production phase. Once R reaches two, the profit split is 85/15% in favour of KRG. The higher the value of the R factor, the more the sharing proportion changes in favour of KRG.

As demonstrated in Fig. 5.16, the contracts also require IOCs to allocate almost five per cent of their profit share to social welfare programmes to ensure lasting benefit to communities affected by petroleum operations (Ministry of Natural Resources 2013a, b, 2015a).

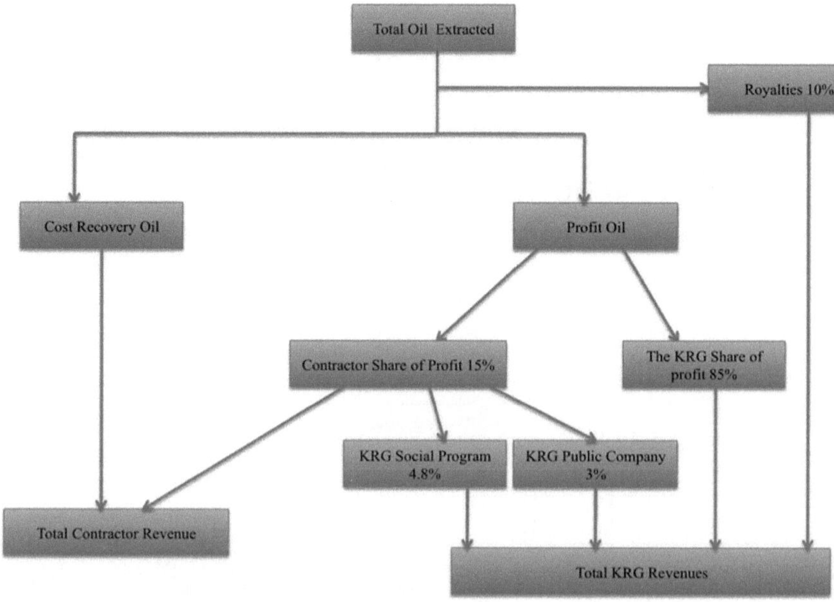

Fig. 5.16 PSC profit sharing in the Kurdistan region. *Source* Based on data from Ministry of Natural Resources (2015a)

The benefits of the Kurdistan region's PSCs seem to outweigh their disadvantages (Jalo 2013). The KRG PSCs have contributed to the rapid growth of the oil and gas sector. They offer incentives to foreign oil companies to invest in the Kurdistan oil and gas sector, while simultaneously favouring KRG if oil prices rise, which in turn increases oil recovery and thus revenues. Moreover, the KRG PSCs force oil companies to protect Kurdistan's environment (Jalo 2013). The oil companies are also engaged in local content development and corporate social responsibility (CSR) programmes through capacity-building payments and their own self-funded CSR projects (Ministry of Natural Resources 2015a).

5.3.6 Governance of the Oil and Gas Industry

Kurdistan's petroleum governance model currently lacks a national oil company (NOC). As illustrated in Fig. 5.17, the Regional Council (RC) and MNR carry out policy-making and regulation respectively (Ministry of Natural Resources 2015a). In fact, MNR handles various regulations, including auditing, health and safety, and economic and social regulations. It also proposes licensing policy and bidding processes to the RC, and supervises petroleum exploration, field development and production, as well as governing all related issues through management committees. MNR also issues guidelines, standards and objectives for petroleum exploration, field development and production phases, and is responsible for petroleum revenue

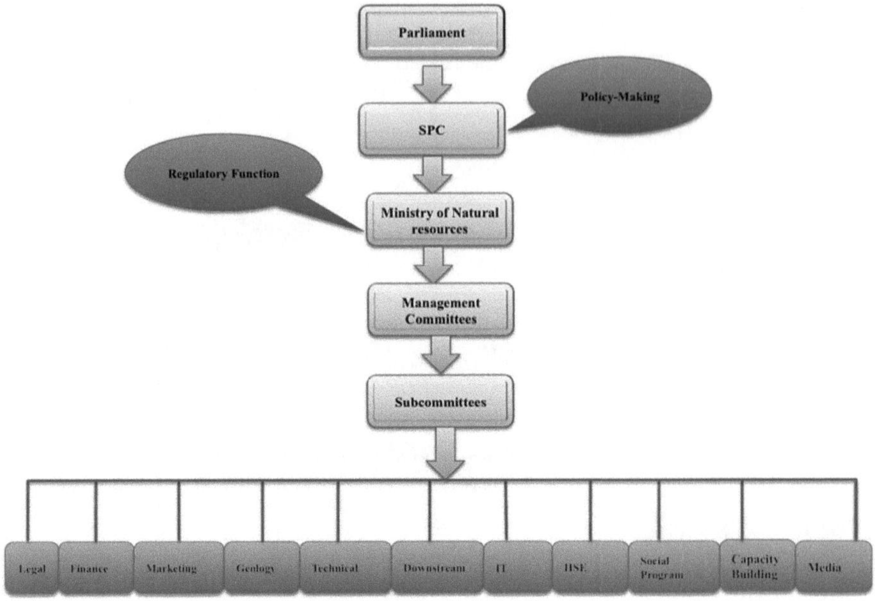

Fig. 5.17 Organisation of MNR. *Source* Author based on data from Ministry of Natural Resources

management, marketing, transport, infrastructure and downstream activities (Ministry of Natural Resources 2015a). Governance of the petroleum sector in the region suffers from inadequate institutions and regulation, and an inefficient localisation strategy. The various public entities appointed by Kurdistan's oil and gas law have not yet been set up. These entities include Kurdistan Exploration and Production Company (KEPCO), Kurdish National Oil Company (KNOC), Kurdish Oil Marketing Organisation (KOMO), and Kurdistan Organisation for Downstream Operations (KODO) (Ministry of Natural Resources 2015c).

The health, safety and environment executives (HSE) lack proper regulations and guidelines, resulting in inefficient monitoring of the environmental performance of petroleum operations, particularly the refinery sector (Whitcomb 2014).

The local content development programme, which aims to localise the oil and gas sector, has been unable to engage the domestic workforce in technical, skilled and managerial positions. Domestic employment is limited to a low-skilled workforce in petroleum activities (see Fig. 5.18). Mismatches in skills supply and demand in the labour market are principal obstacles preventing local graduates from entering oil and gas companies at management level (McIntosh 2014). The inadequately skilled workforce may also be attributed to an ineffective vocational education system that has failed to offer incentives to students. Vocational programmes in the Kurdistan region seem to suffer from poor quality of the training system, which has been unable to make itself more relevant to the labour market, particularly for the extractive industry (UNDP 2012).

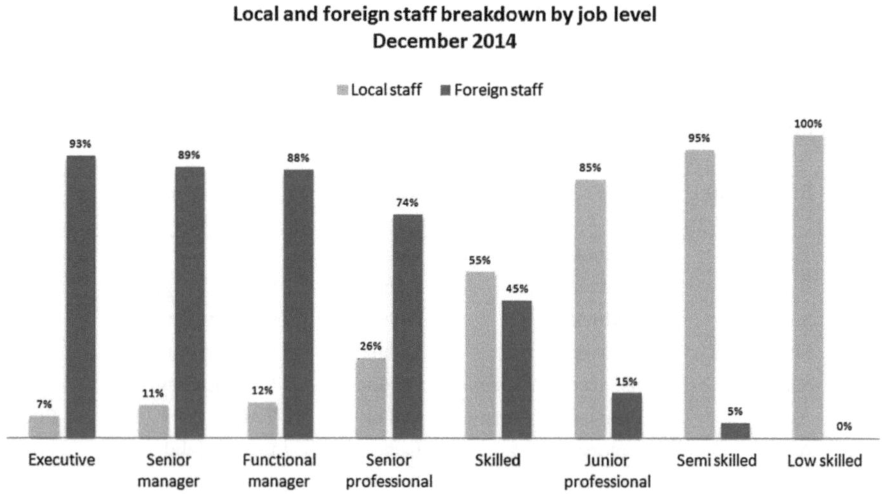

Fig. 5.18 Local and expatriate staff breakdown by level. *Source* McIntosh (2014)

5.4 Conclusions

Oil and gas reserves are important assets for the Kurdistan region. Its oil production capacity has increased and it has reached 637,137 bpd in 2015. This rapid progress has resulted in the region's transformation into a regional and even global economic player.

Despite the opportunities generated by the oil and gas industries, they may also bring risks of high economic dependence on the oil and gas sector due to KRG's failure to diversify the economy away from resource extraction towards productive and sustainable activity, as well as a worsening of the relationship between the people and the government, through economic distortion, lack of transparency and accountability, increased environmental pollution and internal tensions. Furthermore, oil and gas discovery and production lead the people of the Kurdistan region to expect jobs and prosperity. To the extent that these are not realised, the risk of social unrest may increase.

Given this background to the development of the Kurdistan region's economy and oil and gas sector, and the problem of diversifying its economy away from over-reliance on oil revenues, the next chapter discusses in greater depth the political, economic and socio-economic issues surrounding the oil and gas industry in the Kurdistan region through the conduct of semi-structured interview research.

Appendix 1

All licensed fields in Kurdistan region as of 2014 (Oil and Gas Year 2014; Ministry of Natural Resources 2015a).

	Block name	Status	City	Company name (operator)	Year
1	Tawke	**Production**	Duhok	DNO	2008
2	Ber Bahr	Discovery	Duhok	Genel Energy	2009
3	Shaikan	**Production**	Duhok	Gulfkeystone	2007
4	Skeikh Adi	Discovery	Duhok	Gulfkeystone	2009
5	Sarsang	Discovery-on stream	Duhok	HKN Energy	
6	Atrush	Discovery-on stream	Duhok	TAQA	2007
7	Alqush	Exploration	Duhok	ExxonMobil	2011
8	Jebeel Kand	Exploration	Duhok	Turkish Entity	2012
9	AIN SIFNI	Discovery-on stream	Duhok	Hunt Oil	2007
10	Baeshiqa	Exploration	Duhok	ExxonMobil	2011
11	Rovi	Discovery	Duhok	Chevron	2006
12	Sarta	Discovery	Duhok	Chevron	2006
13	Duhok	**Production**	Duhok	DNO	2008
14	Bardarash	**Production**	Erbil	Afren	2008
15	Hawler	**Production**	Erbil	ORYX PETROLEUM	2007
16	Khurmala	**Production**	Erbil	Kar Group (KRG)	
17	Erbil	Discovery-on stream	Erbil	DNO	2008
18	Pirmam	Exploration	Erbil	ExxonMobil	2011
19	Bina Bawi	Discovery-on stream	Erbil	OMV	2008
20	Taq Taq	**Production**	Erbil	Genel Energy & Addax Petroleum	2008
21	Safeen	Exploration	Erbil	Total	2010
22	Harir	Discovery	Erbil	Marathon Oil	2010
23	Shakrok	Exploration	Erbil	Hess	2011
24	Betwata	Exploration	Erbil	ExxonMobil	2011
25	Hindren	Exploration	Erbil	Turkish Entity	2012
26	Choman	Exploration	Erbil	Turkish Entity	2012
27	Dinarta	Exploration	Erbil	Hess	2011
28	Akre Bijeel	Discovery-on stream	Erbil	MOL	2007
29	Qara Hanjeer	Exploration	Kirkuk	ExxonMobil	2011
30	Chemchemal	Discovery-on stream	Sulaimanya	Crescent Petroleum & Dana Gas	
31	Khalakan	Discovery	Sulaimanya	Gas Plus Khalakan	2009
32	Qaladze	Exploration	Sulaimanya	Repsol	2011
33	Piramagrun	Exploration	Sulaimanya	Repsol	2011

(continued)

(continued)

	Block name	Status	City	Company name (operator)	Year
34	Miran	**Production**	Sulaimanya	Genel Energy	2007
35	Bazian	Exploration	Sulaimanya	KNOC	2007
36	Sangaw South	Discovery	Sulaimanya	KNOC	2008
37	Top Khana	Discovery	Sulaimanya	Talisman Energy	2011
38	Kurdamir	Discovery-on stream	Sulaimanya	Talisman Energy	2008
39	Khor Mor	**Production**	Sulaimanya	Crescent Petroleum & Dana Gas	
40	Taza	Discovery	Sulaimanya	Oil Search	2011
41	Pul Khana	Discovery	Sulaimanya	Turkish Entity	2012
42	Garmian	Discovery-on stream	Sulaimanya	Western Zagros	2011
43	Shakal	Discovery	Sulaimanya	Gazprom Neft	2012
44	Chia Surkh	Discovery	Sulaimanya	Genel Energy	2009
45	Qaradagh	Exploration	Sulaimanya	Chevron	2013
46	BARANAN	Exploration	Sulaimanya	Total	2013
47	Arbat	Exploration	Sulaimanya	Turkish Entity	2012
48	Arbat East	Exploration	Sulaimanya	ExxonMobil	2011
49	Halabja	Exploration	Sulaimanya	Gazprom Neft	2013
50	Guwayer	Exploration	Sulaimanya	Kompet Grouop	2013

References

Al-Salhy S, Lyon A (2014) Iraq's Maliki threatens to cut fund if Kurds pipe oil to Turkey. Reuters, 12 January. Available at: http://www.reuters.com/article/us-iraq-kurdistan-oil-idUSBREA0B0AY20140112. Accessed 12 Feb 2016

Barbosa AL (2014) Kurdistan analysis: July 2014. Edinburgh International, London. Available at: http://www.edinburghint.com/fileadmin/user_upload/Documents/EI_-_Kurdistan_Analysis_-_July.pdf. Accessed 11 Feb 2016

British Petroleum (2015) BP statistical review of world energy, 64th edn. Available at: http://www.bp.com/content/dam/bp/pdf/energy-economics/statistical-review-2015/bp-statistical-review-of-world-energy-2015-full-report.pdf. Accessed 11 Feb 2016

Burr JM (2014) Kurdistan oil export: a game changer. American Center for Democracy website commentary. Available at: http://acdemocracy.org/kurdistan-oil-export-a-game-changer/. Accessed 11 Feb 2016

Belkin P, Nichol J, Woehrel S (2013) Europe's energy security: options and challenges to natural gas supply diversification. Congressional Research Service, 7-5700

Coles I, El Gamal R, Zhdannikov D (2015) Kurdish oil deal with Baghdad unravels as tensions rise. Reuters, 13 March. Available at: http://www.reuters.com/article/us-iraq-kurdistan-oil-idUSKBN0M91MQ20150313. Accessed 12 Feb 2016

Crescent Petroleum (2015). Kurdish region. Available at: http://www.crescentpetroleum.com

Chyong C-K, Slavkova L, Tcherneva V (2015) Europe's alternatives to Russian gas. European Council on Foreign Relations website commentary. Available at: http://www.ecfr.eu/article/commentary_europes_alternatives_to_russian_gas311666. Accessed 12 Feb 2016

Dickel R, Hassanzadeh E, Henderson J, Honoré A, El-Katiri L, Pirani S, Rogers H, Stern J Yafimava K (2014) Reducing European dependence on Russian gas. OIES Paper NG92, Oxford Institute for Energy Studies, Oxford

Economic and Finance Committee (2011) Kurdistan budget assessment report. Parliament of the Kurdistan Region, Erbil

Economic and Finance Committee (2012) Kurdistan budget assessment report. Parliament of the Kurdistan Region, Erbil

Economic and Finance Committee (2013) Kurdistan budget assessment report. Parliament of the Kurdistan Region, Erbil

Ekurd Daily (2015) Iraqi Kurdistan spends $3 billion annually on electricity: PM. Ekurd Daily, 16 February. Available at: http://ekurd.net/iraqi-kurdistan-spends-3-bln-annually-on-electricity-pm-2015-02-16. Accessed 12 Feb 2016

Finon D (2011) The EU foreign gas policy of transit corridors: autopsy of the stillborn Nabucco project. OPEC Energy Rev 35(1):47–69

Gray CW (2012) Petrocapitalism in Iraqi Kurdistan: leveraging oil and gas firms in post-war Iraq. CMC Senior Theses, Paper 337, Claremont McKenna College, Claremont, CA. Available at: http://scholarship.claremont.edu/cmc_theses/337. Accessed 12 Feb 2016

Hawramy F, Beaumont P (2014) Iraqi Kurdish forces take Kirkuk as ISIS sets its sights on Baghdad. The Guardian, 12 June. Available at: http://www.theguardian.com/world/2014/jun/12/iraq-isis-kirkuk-baghdad-kurdish-government. Accessed 12 Feb 2016

Heshmati A (2007) Labour market policy options of the Kurdistan Regional Government. Discussion Paper No. 3247, Institute for the Study of Labor, Bonn, Germany

Heshmati A (2012) A review of the regional development strategy for Kurdistan region. Paper presented at the second world Kurdish Congress, Hawler, Iraq, 11–15 October

Holland B (2012) Are Kurdistan's oil contracts constitutional? Petroleum Economist. Available at: http://www.petroleum-economist.com

Invest in Group (2014) Forging ahead. Available at: http://investingroup.org/review/185/forging-ahead-kurdistan/. Accessed 12 Feb 2016

IEA (2012) Iraq energy outlook: world energy outlooks special report. International Energy Agency, Paris

Jalo IP (2013) The rights to explore for and exploit petroleum. What manner of award of rights is best suited for the Iraqi petroleum industry? Centre for energy, petroleum and mineral law and policy, University of Dundee, CAR vol 16

KRSO (2012) Socio-economic monitoring system report. Kurdistan Region Statistics Office, Erbil, Iraq. www.krso.net

KRSO (2014) Kurdistan's population. Kurdistan Region Statistics Office, Erbil, Iraq. www.krso.net

KRSO (2015) Kurdistan's GDP. Kurdistan Region Statistics Office, Erbil, Iraq. www.krso.net

Kurdistan Board of Investment (2015) [website]. Available at: http://www.kurdistaninvestment.org. Accessed 12 Feb 2016

Kurdistan Regional Government (2015) [website]. Available at: http://cabinet.gov.krd

Mackertich DS, Samarrai AI (2015) History of hydrocarbon exploration in the Kurdistan region of Iraq. GeoArabia 20(2):181–220

Ministry of Natural Resources (2013a) Annual financial report. Kurdistan Regional Government, Ministry of Natural Resources, Erbil, Iraq

Ministry of Natural Resources (2013b) Monthly report. Kurdistan Regional Government, Ministry of Natural Resources, Erbil

Ministry of Natural Resources (2014) Production report. Kurdistan Regional Government, Ministry of Natural Resources, Erbil, Iraq. Available at: http://mnr.krg.org

Ministry of Natural Resources (2015a) Kurdistan Regional Government, Ministry of Natural Resources, Erbil, Iraq. Available at: http://mnr.krg.org

Ministry of Natural Resources (2015b) Oil production, export, and consumption report. Kurdistan Regional Government, Ministry of Natural Resources, Erbil, Iraq

Ministry of Natural Resources (2015c) Kurdistan petroleum law. Kurdistan Regional Government, Ministry of Natural Resources, Erbil, Iraq. Available at: http://mnr.krg.org

Ministry of Planning (2015) All capital investment projects report 2007–2014. Kurdistan Regional Government, Ministry of Planning, Erbil

McIntosh I (2014) Who is going to manage the oil and gas industry of the future in Kurdistan. Ministry of Natural Resources. Kurdistan Regional Government, Ministry of Natural Resources, Erbil, Iraq. Available at: http://mnr.krg.org

Paasche TF, Mansurbeg H (2014) Kurdistan Regional Government—Turkish energy relations: a complex partnership. Eurasian Geogr Econ 55(2):111–132

Pflüger F, Duero A (2011) New stability and prospects for Kurdish oil and gas. Eur Energy Rev. Available at: http://europeanenergyreview.com/site/pagina.php?id=3396. Accessed 12 Feb 2016

Rudaw (2014) Kurdistan government blasts Baghdad over budget freezes. Rudaw, 3 March. Available at: http://rudaw.net/english/kurdistan/030320141. Accessed 12 Feb 2016

Shafaq (2014) ISIS and Bagdad siege on Kurdistan hit the economy and drives 100 thousand into unemployment. Shafaq News, 30 September. Available at: http://english.shafaaq.com/index.php/stories/11543-isis-and-baghdad-siege-on-kurdistan-hit-the-economy-and-drives-100-thousand-into-unemployment. Accessed 12 Feb 2016

Sunley EM, Baunsgaard T, Simard D (2003) Revenue from the oil and gas sector: issues and country experience. Fiscal policy formulation and implementation in oil-producing countries, pp 153–183

The Oil and Gas Year (2014) The Kurdistan region of Iraq 2014. United Arab Emirates, Dubai

The Economist (2012) A tortuous triangle. Economist.com, 22 December. Available at: http://www.economist.com/news/middle-east-and-africa/21568760-governments-turkey-iraq-and-iraqi-kurdistan-play-dangerous-game. Accessed 12 Feb 2016

The Economist (2014) A winning hand. Economist.com, 21 June. Available at: http://www.economist.com/news/middle-east-and-africa/21604628-kurds-are-benefiting-mess-iraq-will-independence-follow. Accessed 12 Feb 2016

The Guardian (2014) Iraq government and Kurdish region strike budget and oil export deal. The Guardian, 2 December. Available at: http://www.theguardian.com/world/2014/dec/02/iraq-government-kurdish-budget-oil-export-deal. Accessed 12 Feb 2016

The Jamestown Foundation (2013) Azerbaijan and the southern gas corridor to Europe: implications for U.S. and European energy security. In: Conference report, Washington, DC

UNDP (2009) Budget execution support in the Kurdistan Regional Government. United Nations Development Programme, New York

UNDP (2012) Building the Kurdistan region of Iraq: the socio-economic infrastructure. United Nations Development Programme, New York

UNDP (2013) Corruption and integrity challenges in the public sector of Iraq. United Nations Development Programme, New York

USAID (2008) Kurdistan region: economic development assessment. United States Agency for International Development, Washington, DC

Winrow GM (2013) The southern gas corridor and Turkey's role as an energy transit state and energy hub. Insight Turkey 15(1):145–163

World Bank (2015a) The Kurdistan region of Iraq needs an estimated US$ 1.4 billion this year to stabilize the economy, working paper for a world fee of poverty. Washington, DC: The World Bank. Available at: http://www.worldbank.org. Accessed 12 Feb 2016

Whitcomb A (2014) Oil ministry hints it may clean up its act. Rudaw, 12 November. Available at: http://rudaw.net/english/kurdistan/131120141. Accessed 12 Feb 2016

Yin RK (2009) Case study research: design and methods, 4th edn. Sage Publications, Thousand Oaks, CA

Chapter 6
Management of Oil and Gas Resources in Iraqi-Kurdistan

Abstract Author conducts the qualitative interviewing analysis to provide an in-depth understanding of the challenges facing the newly emerging oil and gas sector in the Kurdistan Region. The results show that multiple issues confront the petroleum industry, including institutional, human resource and geopolitical challenges. Lack of accountability and transparency, ineffective checks and balances, a growing unemployment rate, and ineffective management of public expectation with respect to the new emerging petroleum industry are the major institutional and social challenges facing the Kurdistan Regional Government. The findings also show that high dependency of the Kurdistan region's economy on the hydrocarbon revenues and more importantly ineffective and un-transparent petroleum revenues are the key economic challenges. In addition to the challenges the Kurdistan region's oil and gas sector faces on institutional, human resource, and economic development, there are geopolitical issues, such as being a landlocked region dependent on neighbouring states for market access, the risk of Islamic State attack, and fluctuations in oil prices.

Keywords Accountability and transparency · Checks and balances · Unemployment · Oil price fluctuation · Market access · Security

6.1 Introduction

Chapter 5 discussed the Kurdistan Region's oil and gas sector as its main engine of economic development, and both the internal and external challenges facing this industry were highlighted. A number of key issues remain unresolved between the Federal Government of Iraq (GOI) and the Kurdistan Regional Government (KRG), such as the right to control oil and gas resources and the right to export (and receive revenue) independently. The Kurdistan Region also faces a number of concurrent

K. Abdullah Auzer, *Institutional Design and Capacity to Enhance Effective Governance of Oil and Gas Wealth: The Case of Kurdistan Region*, Perspectives on Development in the Middle East and North Africa (MENA) Region, DOI 10.1007/978-981-10-4518-9_6

challenges, including a financial crisis, the low quality of public services, and chronic shortages in electricity, water and gasoline supplies. In this context, the purpose of this chapter is to develop an improved understanding of the political and socio-economic challenges facing the transformation of the Kurdistan Region's petroleum wealth into other assets that will support sustained economic development.

The analysis presented in this chapter addresses a key research question: how effective is the current governance structure of the oil and gas sector in the Kurdistan Region? In order to answer this question, a case study approach is adopted, which is designed to examine the challenges and opportunities surrounding the development of the oil and gas sector within the Kurdistan Region (Yin 2003).

A semi-structured interview technique was used for data collection, which included pre-formulated questions but was open to new ideas or questions during the interview process (Myers 2013). The twenty interviewees were key stakeholders involved in the oil and gas industry in the region, enabling analysis of the issues facing the development of the hydrocarbon sector from various angles. The target groups were: the Ministry of Natural Resources, the Kurdistan Parliament (Natural Resource Committee), foreign oil and gas companies, domestic suppliers, and the media.

The findings of the previous chapters suggest that institutional strengthening and human capacity-building strategies are of particular importance in understanding the ability of a petroleum resource-rich country or region to deal with a potential 'resources curse'. Therefore, the interview research focused on the following questions relating to the three economic, political and socioeconomic transmission channels of the resource curse:

1. How effective, accountable and transparent are institutions in the Kurdistan Region? What are the challenges and opportunities?
2. How effective and capable are domestic human resources within the Kurdistan Region? What are the existing obstacles and opportunities?
3. What are the challenges currently facing economic development as a whole within the Kurdistan Region?

Having produced transcripts of all the interviews, an inductive thematic analysis strategy was applied to analyse the data (Creswell et al. 2007). The responses were carefully read and coded to look for patterns in the data. In order to verify the accuracy of the emerging codes and categories, they were subjected to independent review. Section 6.2 presents an analysis of the interview data, and Sect. 6.3 summarises the research.

6.2 Interview Analysis

This section presents the interview analysis. Section 6.2.1 presents the five-target groups' perceptions regarding institutional challenges, Sect. 6.2.2 presents the social challenges, and Sect. 6.2.3 presents the economic challenges facing the oil and gas sector in the Kurdistan Region.

6.2.1 Institutional Challenges

In 2007, the Kurdistan Parliament approved a new Oil and Gas Law, which proposed seven separate institutions:

1. The Regional Council (policy-making body)
2. Ministry of Natural Resources (regulatory body)
3. Kurdistan Exploration and Production Company (KEPCO, operational function relating to future fields)
4. Kurdistan National Oil Company (KNOC, operational function with respect to current fields)
5. Kurdistan Oil Marketing Organisation (KOMO, commercial functions)
6. Kurdistan Organisation for Downstream Operations (KODO, operational function regarding downstream activities)
7. Kurdistan Oil and Trust Organisation (KOTO, the petroleum fund).

In this context, the regional Oil and Gas Law suggested a governance model for the oil and gas sector that would separate the policy-making, regulatory, marketing and operational functions. However, the Kurdistan Region currently administers its petroleum resources through two government bodies: the Regional Council (RC), which directs policy, and the Ministry of Natural Resources (MNR), a regulatory body providing oversight and technical expertise as well as engaging in commercial activities. There is no national oil company and none of the suggested entities have so far been established, despite the clear decree of the Oil and Gas Law.

This section examines the effectiveness of oil and gas resource governance in the Kurdistan Region and the potential challenges affecting the current regulatory framework and institutional arrangements. When the respondents were interviewed about the effectiveness of the Kurdistan Region's governance of hydrocarbon wealth, most were of the opinion that the sector has not been managed in an optimal manner. The feedback obtained from the interviewees highlights three main challenges facing the oil and gas sector in the Kurdistan Region: poor institutional design, low institutional quality, and poor safeguards and quality controls.

Poor institutional design

The findings from the interviews reveal that the Kurdistan Region's hydrocarbon industry lacks an effective institutional framework within which to discharge its responsibilities. The interviewees were of the opinion that there are serious issues affecting governance of the region's petroleum sector.

Ineffectiveness of the policy-making body

The participants explained that the Regional Council, as the policy-making body, has not yet been institutionalised. For instance, Interviewee 03 stressed:

> … the entire council is under the influence of the Minister of Natural Resources, since he is the only person with extensive background and expertise in the oil and gas industry. Furthermore, one cannot exclude incidents of conflict of interest, as the person who is head of the regulatory body, in this case the Minister of Natural Resources, is also an effective and even an influential member of the Regional Council.

Interviewee 17 elaborated more explicitly:

> …the Regional Council is not working as an independent agency and lacks experts and a technical body to advise the council in terms of political, technical, managerial and economic subjects. Therefore, it is not able to make effective petroleum policy and oversee the Ministry of Natural Resources.

In the same vein, Interviewee 05 commented:

> The Regional Council lacks an institutional structure which will enable the policy makers to formulate prudent policy based on appropriate and effective 'checks and balances' for the petroleum industry, as well as to monitor the implementation of the formulated policy by the Ministry of Natural Resources.

Delay in the establishment of entities in compliance with the region's 2007 Oil and Gas Law

In addition, the majority of interviewees stated that inadequate agencies impact on the performance of Kurdistan's oil and gas sector. Interviewee 02 said:

> I believe that the KRG will not be able to effectively exercise authority over the oil and gas industry in the Kurdistan region unless it possesses powerful, competent and state-owned upstream and downstream companies.

Furthermore, Interviewee 09 emphasised that:

> The petroleum sector suffers from an ineffective institutional framework due to a lack of state-owned companies and other agencies, which has also necessitated endorsement of the concept of "separation roles" as a fundamental tool to build a healthy business culture to prevent conflict of interest.

Interviewee 04 expanded on this point:

> The establishment of the stated-owned oil company is very necessary. The Ministry of Natural Resources currently has two potentially conflicting roles: that of regulator and secondly of joint-venture partner (through the government share). The Management Committee (MC) reflects the JV aspect, but the regulatory role is not respected and the MC Chairmen often also try to act in this capacity, even getting into a conflict of interest situation.

Political and powerful elite group interference

The third reason why most respondents did not believe that the oil and gas sector is successful and effective is because of the politicisation of the regulatory and oversight function of the Ministry of Natural Resources. Interviewee 10 said:

> …the dominant political parties and their close network have positioned themselves to profit from the sector by functioning as a petroleum industry gatekeeper for foreign companies and lining up private companies to take advantage of contracts, particularly in the refinery sector and the area of logistics.

Interviewee 05 stated:

> …the tendering system is not transparent in the Ministry of Natural Resources because of political interference and elite groups. Therefore, business activities are based not on competiveness but on monopoly for a number of companies, which many claim belong to powerful political parties and their associates.

Overall, the interviewees revealed that the current organisational structure of the oil and gas sector endorsed by the Ministry of Natural Resources fails to maximise the benefits and meet the challenges of providing an appropriate petroleum management system and procedures, as required by the Oil and Gas Law. A weak policy-making body, namely the Regional Council, and the absence of a national oil company has created a ministry-dominant model of petroleum governance. In this context, the government ministry carries out all policy-making, regulatory and monitoring functions. Moreover, the lack of effective institutional building poses a severe challenge for transparency and accountability in the Kurdistan Region's hydrocarbon sector, raising doubts among the public. Such observations are largely in agreement with results reported in the literature (e.g. Heller and Marcel 2012; Lahn et al. 2009; Thurber et al. 2010, 2011; Al-Kasim et al. 2008, 2013; Al-Kasim 2006; Barma et al. 2012). These authors conclude that a well-designed institutional framework is crucial for the effective management and oversight of all phases of the oil and gas sector, thereby enhancing the performance of the petroleum industry. In addition to the importance of strong institutions, the findings highlight the current structure of state ownership without accountability, which contribute to the poor quality of institutions and negative economic outcomes for the oil and gas sector (Luong and Weinthak 2006). Under this ownership structure and as a result of the lack of a stated-owned oil company, the Kurdistan Regional Government has granted significant managerial and operational control to international oil companies (IOCs) through production-sharing agreements. The IOCs are the able to play a major role in the management of petroleum wealth by exercising control over operations, investment decisions and production levels (Stevens 2008).

Given this explanation, it can be concluded that inadequate and ineffective institutional design of the petroleum sector has resulted in the poor management of oil and gas wealth in the Kurdistan Region.

Low institutional capacity

Participants indicated a number of institutional challenges facing the Kurdistan Region, stemming from weak institutions relating to governance of the oil and gas sector. These factors include excessive bureaucracy, low technical capability, a lack of accurate and consistent data publishing, and interference by political and elite groups.

Excessive bureaucracy

Almost all participants agreed that excessive bureaucracy is troublesome because it lays the ground for a decline in foreign investment and, in turn, to poor economic performance. Interviewee 11 said:

> The Kurdistan Region's oil and gas sector suffers from the bureaucratic/administrative-centred system. There is huge bureaucracy here; instead of dealing with one person it is with multiple people over one issue. There is not a single entity to deal with it – namely, long routine and bureaucracy makes the environment less attractive for foreign investors…

Interviewee 08 also said:

> The oil sector is suffering from the red tape of bureaucracy. In Nigeria, which is a difficult place to work, you find just one person whom you deal with, you know with whom you deal; but here, the investors or foreign companies deal with many different people. This is costly and time-consuming.

Interviewee 15 finally stated:

> Excessive bureaucracy is the major impediment to our activities. Government bureaucracy and corruption increase the cost of doing business. The firms use bribes to lower their regulatory burden in the Kurdistan Region.

In short, the high level of bureaucracy is a major obstacle facing oil companies in the Kurdistan Region. Investment regulation is friendly and attractive and offers incentives to investors, but when they arrive and face a great deal of paperwork and bureaucracy they become frustrated. The long, cumbersome bureaucracy is due to a lack of coordination and collaboration between the various departments in the ministry. The problems cited by the participants are consistent with the literature on institutional challenges in oil- and gas-rich countries. For example, a study conducted by Eifert et al. (2002) sheds light on the fact that bureaucratic red tape is an obstacle to effective oil governance in oil-dependent countries, such as the Gulf States. An over-expanded bureaucracy leads to rent-seeking activities that distort economic development (Karl 2007).

Lack of technical capability

The majority of participants agreed that a basic obstacle to the effective governance of the oil and gas industry is the shortage of indigenous technological capacity. Interviewee 14 explained:

> There is insufficient expertise in the Ministry, and the systems and processes for effective oversight are largely non-existent. Much ad hoc control rests with the Management Committee chairmen, who are not necessarily skilled to address matters in all areas – legal, technical, commercial, human resources, environment, etc.

Interviewee 02 also indicated:

> The ministry suffers from a lack of efficient capacity to deal with issues. Owing to the lack of competence in the Ministry, certain issues become more complicated and time-consuming rather being resolved.

Other interviewees stated that Kurdistan's oil and gas sector lacks a research and development entity, which is a cornerstone for the development of knowledge and technology in respect of various petroleum activities. Interviewee 09 stated:

> Technical support, consultation, research and technology for the local petroleum industry are provided outside the region because the establishment of an institute for research has been delayed.

Another participant 13 added:

> Setting up an institute for research is the key institution for research and development within the petroleum sector. This can support assessments of industries and technological development in the Kurdistan Region and joint scientific projects with organisations outside the region.

In addition, Interviewee 07 said:

> Without a sound energy institute the Kurdistan region will be deprived of a framework for international research which could potentially lay the foundations for relevant, research-based study programmes with the potential to support interdepartmental research programmes. It would be very difficult to identify a new technological and conceptual solution that boosts oil and gas extraction in the Kurdistan region without having an integrative centre of excellence.

Most interview participants shared a common view that the weak regulatory body impacts on a variety of areas, including timeliness, predictability and cost-effectiveness, resulting in duplication of effort and poor innovation. In this context, the weak regulatory system makes corruption possible in all petroleum activities from exploration to the refinery sector (McPherson and MacSearraigh 2007).

Accurate and consistent data publishing

Transparency is perceived to be lacking as a result of failure to publish production and other significant data regularly. Since 2007, all contracts awarded by the Ministry of Natural Resources (MNR) have been published on the Ministry's website, but contracts signed with international oil companies have not been published since 2012. Interviewee 05 stated:

> The Kurdistan Region Government has signed a contract for fifty years with Turkey. It is acceptable not to publish these contracts until they are accomplished because of commercial or national interests, but it is not normal for them to remain unpublished. I am sure that the Federal Government has one version of these contracts, and why the Kurdistan Parliament should not have them I don't know. For example, the argument that the MNR made regarding the confidentiality of the contract between Kurdistan and Turkey for fifty years is not valid because there is no need to be confidential; the same contract that the Iraqi government signed in 1973 for seventy years with Turkey was published.

Given this explanation, keeping contracts secret simply reflects the fact that the MNR has not worked effectively with the Natural Resources and Economic and Financial Committees of parliament. This has reduced trust between parliament and government (MNR), and thus it has lost parliamentary support for its petroleum policy. Interviewee 18 also confirmed that non-disclosure of information and inconsistent reporting has resulted in increased doubt about transparent management of the petroleum sector:

> …we see disclosure of information and reporting is increasing but we still see inconsistency in the timing of it. For instance, you see some financial reports come out from the ministry that are useful but on an inconsistent basis; there are some production reports from the ministry, but again inconsistent. We see some product-sharing contracts (PSCs) published, then publication halts – once again inconsistent.

However, some interviewees believed that the Kurdistan Region's primary purpose is to increase the transparency and accountability of hydrocarbon management. Interviewee 01 said:

> …There are few petroleum-rich countries that publish oil contracts; for example Malaysia, Indonesia and Egypt have never published contracts… in contrast to the KRG, the Federal Government has not published any contracts. The reality is that we try to be transparent.

Interviewee 01 explained further: "In effect, there are two major reasons for unpublished contracts: first because of commercial issues, and second the interests of oil companies." In addition, a number of participants explained that the Regional Council highlighted details of unpublished contracts for members of parliament (MPs). Interviewee 04 also confirmed that the government has attempted to make the petroleum sector transparent and accountable:

> But some oil contracts, such as the contracts with Turkey, have not been published because of national interests. The Regional Council stated that it is ready to inform MPs of the details at the Council of Ministries, but MPs did not follow it up. Even the prime minister announced his readiness to discuss all details with all heads of faction parties, but they have not yet responded and do not show any interest in the Prime Minister's suggestion.

In addition, the lack of accurate, detailed data has not only had implications with respect to making the governance of hydrocarbon wealth less transparent, but has also led to political conflict between the KRG and Federal Government in Baghdad. For example, Interviewee 13 explained: "Some of the problems the KRG has with Baghdad come down to accurate oil- and gas-related data." Interviewee 20 also stated: "The financial reports are not in the detailed information; however, the only published report is regarding up to the 2013 fiscal year."

In short, the greater concern is that there seems to be a lack of accountability and transparency to the people of Kurdistan. The challenge is to make the Ministry of Natural Resources more effective, specifically as a regulator. One of the most vital mechanisms in successful petroleum governance is the accountability practice of the regulatory agency involved (McPherson 2003).

Poor safeguards and quality controls

Most participants explained that the absence of an effective system of checks and balances has led to poor governance of the oil and gas sector in the Kurdistan Region. The low capability of Parliament and the Supreme Auditor Council, and a lack of international organisations such as the Extractive Industries Transparency Initiative (EITI) are the main factors that interviewees linked to poor oversight of petroleum activities.

Interviewees stated that the low capability of MPs in terms of an understanding the oil and gas industry is the main reason behind the weak monitoring of the oil and gas sector. For example, Interviewee 02 said:

> …a lot of questions we see from the MPs are in one direction and show their understanding of the industry could be better, and this can lead to misunderstanding that leads to mistrust. What would be better is for the MNR to provide better details so that the public can be more educated about the whole oil industry, so they can understand the questions and be aware of the purpose of the questions.

Moreover, the participants also indicated that none of the members of the Natural Resource Committee of Parliament are familiar with the oil and gas industry. Interviewee 11 explained:

> I don't know anyone who works in the Natural Resource Committee of Parliament with any expertise in the oil and gas field. This committee is not capable of making any dramatic changes; so you cannot conduct quality control on something if you are yourself lacking the background expertise.

Most participants argued that published revenue and other data are not valid because a third party has not audited them. The data need to be audited by the Auditor General's Office. With regard to international organisations, most interviewees indicated that the engagement of international organisations such as EITI and the World Bank could play a key role in improving quality control in the petroleum sector.

In short, the Natural Resource Committee of the Kurdistan Parliament is the only public authority with a constitutional mandate to debate and influence petroleum policy through the law-making process. It oversees the implementation of oil and gas policy by the Ministry of Natural Resources (Kurdistan Regional Government 2015). The effective oversight function of parliament is crucial to the transparency and accountability of oil and gas governance (Myers and Mohammed 2012; World Bank 2010; Lindner 2014). Specialised knowledge of members of the Natural Resource Committee would increase their ability to influence policy decisions and oversee performance in the petroleum industry (Acosta 2010).

6.2.2 Social Challenges

The majority of interviewees revealed that governance of the oil and gas industry in the Kurdistan Region has faced two key social challenges: unemployment, and ineffective management of public expectations.

Unemployment

According to the interviewees, youth unemployment is a major social challenge for the region. All participants agreed that the KRG has failed to design polices to deal with the issue of increasing public employment by accelerating job growth in the private sector. One of the most critical problems highlighted in the interviews regarding the policy of job creation for young people is the increasing appeal of public sector employment. For example, Interviewee 08 said:

> The public sector is heavily preferred owing to the benefits it offers: the employer has little or no control over the employee, high public sector salaries and the low match between the skills of public sector employees and the work roles they perform.

Interviewee 17 stated: "Working hours in the public sector are shorter than in the private sector, and government jobs are much safer than those in the private sector." Interviewee 14 also indicated:

> The recruitment system is not based on a merit system. The selection of applicants for employment is influenced by political interference, elite groups and personal connections rather than skills and competence…

In addition to the attractiveness of the public sector, nearly all interviewees highlighted that a lack of incentives and subsidies to increase private employment makes working in the private sector less attractive. Interviewee 06 said: "There are not better wage expectations for local job seekers moving to the private sector." Interviewee 19 said: "There is little interest in proactive employment development in the industry by offering the private sector financial and other incentives to employ the local workforce."

Another major bottleneck to youth employment cited by interviewees is mismatches in education and skills; namely, a serious mismatch between the skills young people bring with them when they leave the education system and those that are sought in labour markets in the Kurdistan Region. In addition, many interview participants identified the low quality of the educational and vocational system as a principal obstacle to young job seekers in labour markets. For instance, Interviewee 09 said: "We have difficulties in finding suitably skilled workers. There are local skill gaps and deficits with respect to the petroleum sector." Interviewee 13 stated:

> From my experience, the big challenge facing human resource development is qualification mismatches and the quality of education and skills. Employers have a great struggle to find suitable candidates with the proper education and skills required by the job.

These results are consistent with the political economy literature (e.g. Gavin 1993; Auty 1993, 1997; Gelb 1989, 2011; Robinson et al. 2006), which shows that a common way of distributing rents in petroleum exports is to increase public spending on civil service wages. Furthermore, inefficient allocation of windfalls has hampered economic development polices, such as domestic capacity building with appropriate quality and capabilities comparable to international standards. This context makes it difficult to enhance indigenous participation in petroleum activities, use local technology, award more contracts to indigenous firms, or stimulate joint venture arrangements between indigenous and foreign companies.

Ineffective management of public expectations

Since the newly emerging oil and gas sector in the Kurdistan Region is a source of substantial revenue, the public has been keeping an eye on this industry. The local community appears already to have high expectations relating to the discovery and production of oil. Kurds in the region hope that oil exploitation will improve their quality of life. The general public also expects that oil will increase national prosperity as the KRG invests in roads, power plants, education, health and other socio-economic infrastructure. They expect to see the end of dependency on the federal government and to gain economic sovereignty. Most participants made it clear that poor communication, a low level of public involvement, and a lack of reliable and timely information are significant contributing factors to the poor performance of the government ministry in relation to managing and meeting domestic expectations. For example, Interviewee 20 said: "The Ministry does not have a clear strategy as to how to communicate effectively with the general public. The government promised what it could not realistically deliver." Interviewee 07 stated:

> The Kurds were excited about the huge discovery of oil in Kurdistan and they are expecting to take a share of the proceeds. They have not been educated about things like PSC contracts and other arrangements. As a result, they rely on politicians and media and other individuals for information – sometimes incorrect information.

Interviewee 11 was of the same opinion:

> The government's failure to educate people about the oil and gas industry, how it operates, what is possible and what is not. There is a wrong understanding of the industry among people. They think that this new sector can assure a job for everyone who expresses interest. In this context, there is great room for the media to mislead the Kurds.

The respondents also reported that local participation in petroleum activities is crucial to the effective management of public expectations. For example, Interviewee 10 stated:

> The Kurdish communities feel like outsiders in the oil and gas industry because their involvement in petroleum activities is limited to low-skilled positions. Locally educated people are questioning how training opportunities and scholarships programmes are awarded.

In addition to stepping up technical support and training for local capacity building, the involvement of professionals and civil social organisations in petroleum sector development is crucial in helping shape public expectations. Interviewee 16 said:

> Experts on oil and gas and civil society organisations have not been involved in setting petroleum policies. A lack of dialogue between the government and local experts, as well as civil society, about petroleum policies and current development in the oil and gas sectors leads to ongoing crises in the region, such as fuel-price protests and electricity shortages.

Failure to provide reliable and timely information to the public has also contributed to a failure to reinforce trust between the KRG and its citizens. In this context, the media have furthered public confusion by reporting on issues on which they lack in-depth knowledge. Interviewee 20 said: "The media often take the figures out of the original publications of various organisations and report them as if they come from a realistic base rather than speculative one." Most participants indicated that the government plays a major role in providing true information to the public, thereby building realistic expectations. Interviewee 08 explained:

> The oil and gas sector is politically sensitive in the region; every piece of information on the sector attracts media attention, therefore there is a need to manage public expectations effectively. The government has failed to manage public expectations by providing the media with accurate information relating to the oil and gas industry.

Interviewee 03 added:

> The public has not been educated about the various challenges around the oil and gas industry, such as uncertainty in reserve estimations, production rates, timing and magnitude of revenues. For example, people are aware of cheaper gasoline in the rest of Iraq or neighbouring countries, but they do not understand that it is the result of the KRG's subsidisation policy which is not quite sustainable in the long term.

These findings are consistent with research conducted by Marcel (2015) , who argues that community engagement and effective communication with the public are crucial to establishing trust within the public. Furthermore, failure to manage the expectations of local communities efficiently has had negative consequences: people's resentment of the government is rising. This has also seriously impacted on the popularity of the political leadership in the Kurdistan Region and has created mistrust within society.

6.2.3 Economic Challenges

This section examines the major problems affecting the Kurdistan Region's economy. The interviewees indicated that non-transparent revenue collection and spending, lack of economic links with the rest of the economy and geopolitical issues are significant obstacles to economic development of the Kurdistan Region.

Non-transparent revenue collection and spending

Most participants explained that the KRG has failed to improve petroleum revenue transparency. Interviewee 19 said:

> [lak]Petroleum revenue management has been an ongoing issue in the region since the Ministry of Natural Resources has received oil and gas revenues and control over

> expenditures. Such approaches have led to theft, fraud, abuse, and have enhanced political instability and thereby economic stagnation in the region.

Interviewee 05 said:

> Transparent and accountable management of oil and gas revenue flow has been hindered by delays in the establishment of a petroleum fund. Consequently, an absence of internal controls, supervision and transparency has enabled not only mismanagement of oil and gas revenues but also high profits for political and elite groups.

The second reason for ineffective petroleum revenue management is a lack of financial transparency: the government ministry has not disclosed detailed data with respect to how much revenue it has received from its extractive industry and how it spends the revenues. For instance, Interviewee 06 stated:

> A lack of detailed key oil revenue data leaves citizens unable to hold their government accountable for the revenues generated and the use of them –namely, whether their petroleum revenues are being managed responsibly.

Interviewee 16 agreed: "The financial reports have not been consistently disclosed on the website in a way that the public can easily find them, and the revenue needs to be reported as disaggregated figures."

Most respondents agreed with the statement that an ineffective Auditor General's Office has been a contributing factor to the lack of transparency and accountability of petroleum revenue management, thereby increasing abuse of petroleum revenues and corrupt activities in the Kurdistan Region. Interviewee 00 explained: "The Auditor General's Office has engaged ineffectively in the oversight of the petroleum sector." Interviewee 03 said: "The failure of the Auditor General's Office to oversee and audit petroleum sector processes, as well as the production and dissemination of audit reports, has reduced petroleum revenue accountability and transparency."

In short, government accountability and transparency has been restrained by a lack of detailed and consistent publication of revenue figures and a poor monitoring and auditing system. In this context, the exploitation of petroleum revenues has resulted in poverty, inequality, corruption and conflict rather than fostering economic growth and social prosperity (Kolstad and Wiig 2008; Ross 2008).

Lack of economic links

Most participants indicated that Kurdistan's economy is plagued by high dependency on the oil and gas sector. Furthermore, the government has been unable to generate further benefits from petroleum resources for the rest of the economy. The interviewees identified poor local content plans and failure to form other economic links as the main reasons for the oil-dependent economy.

The participants revealed that the Kurdistan Region's oil and gas industry local content policy has not yet achieved significant success. It has failed to enhance the local participation, awards to indigenous firms and the stimulation of joint venture arrangements between local and foreign oil firms. Interviewee 07 said: "The current oil and gas industry local content policy has resulted in the lower participation of indigenous small to medium-sized firms within the industry."

Other interviewees, on the other hand, pointed to the fact that many indigenous service companies are uncompetitive with regard to prescribed contracts or services, as such firms demonstrate insufficient ownership of equipment, Kurdish personnel and the capacity to perform such operations. Therefore, the participation of local servicing firms is constrained to facilities such as catering and basic civil services. Interviewee 18 said:

> Local servicing companies are less competitive in the oil and gas sector. The government has been unable to encourage foreign extractive investors to foster deep links with the local economy through training and guidance of local firms, effective corporate social responsibility (CSR) programmes, and engagement in local capacity building.

The majority of interviewees indicated that the problem is rather a low absorption capacity by domestic economic entrepreneurs that hinders the formation of economic links. Interviewee 05 further explained: "The real problem is political and ruling elite interference in industrial policies for local content development." Interviewee 04 agreed: "Non-transparent and corrupt institutions, and thereby poor governance of the oil and gas industry, has led to deeply rooted rent-seeking activities in local content." Interviewee 13 said:

> Ruling elites are interested in supporting and engaging with domestic economic entrepreneurs who have already captured local content markets, so they obtain rents from these firms to maintain their political power.

In addition to the lack of economic links within the oil and gas industry promoting the local supply industry, the poor contribution of oil and gas revenues to other economic sectors that will boost employment and income generation, such as agriculture, industry and services, is a major barrier to long-term economic development in the Kurdistan Region. Interviewee 15 said:

> Reasons for the lack of linkage formation so far from the petroleum sector are numerous. Most importantly, economic development policies lack the specific targets, monitoring mechanisms, incentives and sanctions needed to reach the overall development objectives relating to other economic sectors, in particular the agricultural sector.

Similarly, interviewee 10 explained that "the present institutional environment offers rent-seeking opportunities for ruling elites rather than productive activities within other economic sectors".

In short, the participants' responses show that the Kurdistan Region's economy is based on petroleum revenues, which has converted the region into a 'rentier state'. Poor institutional quality and human resource capacity are the main reasons behind the inefficient and petroleum revenue-driven economy. These results are in line with the political economy literature discussed in detail in Chap. 2.

6.2.4 Geopolitical Challenges

Interviewees discussed the further challenge of geopolitical issues facing the Kurdistan Region's oil and gas sector, which are a major source of uncertainty. These include the high economic dependency of the KRG on the federal budget, being a landlocked region, the risk of Islamic State attack, and fluctuations in oil prices.

All interviews agreed that the Kurdistan Region has become worryingly over-reliant on the 17% share of the federal budget allotted by Baghdad. Interviewee 17 stated: "The Federal Government has used the region's ongoing financial dependency—namely Kurdistan's budgetary instalments from the federal budget—as a tool to rein in the Kurds' economic development ambitions." Interviewee 03 said: "Control over oil revenues gives Baghdad financial leverage to use against the Kurds. In 2014 it displayed this leverage by refusing to pay the Kurdistan Region's share of the national budget." Interviewee 05 pointed to Kurdistan's disputes with Baghdad:

> The economic sanction that has been imposed on the Kurds since 2014, as a result of the political tension between Erbil and Baghdad, sheds light on the fact that the KRG relies heavily on its budget share, which has made the region vulnerable to economic bankruptcy.

All interviewees agreed that the financial dispute between Erbil and Baghdad has had the effect of delaying payments to oil and gas companies operating in the Kurdistan Region. This has posed a major threat to investment plans in the oil and gas industry and other economic sectors.

Another key geopolitical issue outlined by interviewees is linked to petroleum exports. They explained that the Kurdistan Region is a landlocked region, which raises issues of infrastructural dependency in exporting its oil and gas. Interviewee 09 said: "The Kurdistan Region relies on Turkey for oil and gas exports. Expansion of pipeline exports is the lifeblood of the Kurdistan economy. The future of landlocked Kurdistan depends on political relations between Ankara and Erbil." Interviewee 01 stated: "The Kurds are landlocked and therefore always dependent on others, not only for gas and oil exports but also for imports. Turkey is the best option for the Kurdistan Regional Government because the infrastructure is more or less ready." This interviewee's account shows that the Kurdistan Region needs cooperation with its neighbours to access world markets and thereby guarantee the development of the petroleum industry through further foreign investment in the hydrocarbon sector.

Another critical challenge that has affected the oil and gas industry is security. Interviewee 02 said: "Islamic State attacks have affected operating activities in blocks which are close to the war." Interviewee 16 further explained that:

> After the worsening security threat, oil companies operating in the Kurdistan Region have had varied responses to the current security challenges. Some of them have evacuated non-essential staff and others have stopped part of their activities.

In short, all participants highlighted that security plays a vital role in the ongoing development of the petroleum sector. Security matters for optimal and long-term investments and employment development, which have been hit particularly badly since companies are limiting their commitment as a result of insurgent attacks.

All participants explained that the collapse in oil prices has affected the region's economy, which has worsened the financial crisis in the Kurdistan Region. Interviewee 10 said: "The KRG has started to sell its crude oil independently after financial sanctions were imposed by the central government. However, the KRG has been unable to pay civil servant salaries because of the sharp fall in oil prices." Interviewee 04 said: "The oil revenue-based economy of the region has been hit hard by the decline in oil prices; this has had a huge impact on projects, and most investment projects in the region have been halted in all sectors."

The participants' responses shed light on the fact that the Kurdistan Region's dependency on oil revenues is a major challenge because of poor financial policy in dealing with fluctuating oil prices. For example, Interviewee 18 explained: "The Kurdish security forces have been able to prevent Islamic State attacks and the KRG has also increased oil production capacity but it is suffering a financial crisis." Under these conditions, the collapse in oil prices has resulted in poor welfare and thereby social conflict due to the low quality of institutions. These findings are consistent with the political economy literature discussed in detail in Chap. 2.

6.3 Summary

This chapter has presented an analysis of interview data to provide an in-depth understanding of the challenges facing the newly emerging oil and gas sector in the Kurdistan Region. The results show that multiple issues confront the petroleum industry, including institutional, human resource and geopolitical challenges.

The main institutional issues contributed to the poor design and capacity. Lack of effective checks and balances with in the system has resulted in the low quality of institutions involving in the oil and gas sector. The findings indicate that a growing unemployment rate and ineffective management of public expectation with respect to the new emerging petroleum industry are the major social challenges facing the Kurdistan Regional Government.

High dependency of the Kurdistan region's economy on the hydrocarbon revenues and more importantly ineffective and un-transparent petroleum revenues are the key challenges on the path of the region's economic development. Furthermore, geopolitical factors have impacted on the performance of the oil and gas sector through the high economic dependency of the KRG on the federal budget, the problems of being a landlocked region dependent on neighbouring states for market access, the risk of Islamic State attack, and fluctuations in oil prices.

Appendix A

Interview questions

This interview research includes examining the details that surround the challenges and opportunities associated with development of the oil and gas sector within the Kurdistan Region. It is a semi-structured interview and the target groups who are involved in the oil and gas sector include:

- **Ministry of Natural Resources**: According to Kurdistan's petroleum law, the Ministry shall **oversee** and **regulate** its Petroleum Operation. The responsibilities of the Ministry include the **formulation**, **implementation**, **supervision**, **inspection**, **auditing** and **enforcement of all petroleum operations** by all persons and all activities relating thereto. This includes the **marketing of petroleum**, and **negotiating**, agreeing and executing all authorisations, including **petroleum contracts**, entered into by the Regional Government as well as for amending the terms of any authorisation to ensure that the petroleum operations are carried out for the benefit of the people of the Region and Iraq.
- **Kurdistan Parliament (Natural Resource Committee)**: The committee as the highest legal authority is responsible for overseeing and monitoring the Ministry of Natural Resource's activities.
- **The Foreign Oil and Gas Companies**: The foreign investors involved in the Kurdistan's petroleum industry.
- **Domestic Suppliers**: The indigenous investors involved in the Kurdistan's petroleum industry.
- **Media**: The Kurdish journalists who work within Kurdistan's media
- My research to date suggests that the economic, political, and socio-economic areas are of particularly importance in understanding the ability of a petroleum resource rich country or region to deal with a potential resources curse. Therefore the following are critical and generic questions related to the three areas contained in the interview research:
- How effective, accountable, and transparent are the institutions in the Kurdistan Region? What are the challenges and opportunities?
- How effective and capable are the domestic human resources within the Kurdistan Region? What are the existing obstacles and opportunities?
- What are the challenges currently facing the economic development as a whole within the Kurdistan Region?
- **Interview Questions**
- **Ministry of Natural of Resources**

  - How effective, accountable, and transparent are the institutions within the Kurdistan Region? What are the apparent challenges and opportunities?
  - How effective and transparent is the licensing process? (Announcement—application—negotiation; negotiation contract team)

  - How effective is the monitoring and supervision system currently in place? (Exploration, production activities, auditing, environment and safety issues). Is there adequate skill and sufficient technical knowledge in place?
  - How efficient is the data management solution and is an IT system in place?

- (This is based upon a consistent data model, a reference/administrative database for most important types of data, modern data storage media and well-maintained hardware, allowing efficient use of GIS and other petroleum management software systems).

  - What are your thoughts about the establishment of the Kurdistan National Oil Company? Is it necessary? If yes, what role will this national oil company play in the petroleum sector? Which operational mechanisms should be considered?
  - How effective is the Foreign Investment law within Kurdistan in terms of attracting the foreign investors? What are the current obstacles?
  - How is the petroleum sector affected due to increasing instability on Kurdistan's borders?
  - To what extent has domestic employment been increasing within oil and gas sector? How effective is Kurdistan's domestic training program? What are the prescient obstacles?

- **The Natural Resource Committee of Parliament**

  - How capable and independent are the policymaking and regulatory authorities within Kurdistan's oil and gas sector? (Political and elite group interest as opposed to the national interest).
  - How effective and transparent is the licensing process? (Block announcement, application, and negotiation)
  - How effective is the supervision of the committee on the policy-making process and the implement of the petroleum polices? (Setting petroleum policy, auditing and environment issues)
  - What are the related obstacles? (Lack of cooperation, and committee's technical capability)
  - What are your thoughts about the establishment of the Kurdistan National Company? Is it necessary? If so, what role will the national oil company play in the emerging petroleum sector? Which operational mechanisms should be considered?
  - How effective are the investment regulations law within Kurdistan to attract eager foreign investors to the petroleum and other industries?
  - To what extent has the domestic employment been increasing? How effective is Kurdish domestic training program? What are its present obstacles?

- **International Oil Companies**

  - How effective are the regulatory institutions in Kurdistan?

- (Less bureaucratic, competent public institutions, the tax institution)

- How effective is the Foreign Investment Law within Kurdistan in attracting the interested foreign investors?
- What are the obstacles that exist and the present opportunities available to foreign investors who are tending to enter in Kurdistan with greater frequency?
- How has the petroleum sector been affected by the increasing instability on the Kurdistan's borders? (A decline in investment or change of shareholders in PSCs contracts)
- To what extent has domestic employment been increasing? How effective is Kurdistan's domestic training program? What obstacles exist in its implementation?

• **Domestic Suppliers (local content)**

- How transparent is the awarding of the subcontracts?
- How supportive is the government in encouraging its indigenous industries?

• (Financial support)

- How efficient is the related legislation?
- How competitive are Kurdish companies when compared to the foreign companies working within Kurdistan?
- What are the obstacles and opportunities?

• **Media: Kurdish Journalists**

- How easy is for the media to access the relevant data associated with the oil and gas sector?
- How greatly does the public trust that the petroleum wealth will bring increased welfare and blessing for them?
- To what extent has the government been successful in developing Kurdish talent within the hydrocarbon industry and also within its domestic companies?
- What are the obstacles that exist vis-a-vis domestic capacity building?
- How has the petroleum sector been affected after a marked increase of instability along the borders of Kurdistan?

References

Acosta AM (2010) Pilot programme to strengthen parliamentary oversight of the extractive sector in Ghana and Tanzania: comparative overview of country baselines and M&E framework. Institute of Development Studies, Brighton

Al-Kasim F (2006) Managing petroleum resources: the "Norwegian model" in a broad perspective, vol 30. Oxford Institute for Energy Studies

Al-Kasim F, Søreide T, Williams A (2008) Grand corruption in the regulation of oil. U4 Anti-Corruption Resource Centre, Bergen, Norway

Al-Kasim F, Søreide T, Williams A (2013) Corruption and reduced oil production: an additional resource curse factor? Energ Policy 54:137–147
Auty RM (1993) Sustaining development in mineral economies: the resource curse book. Routledge, London
Auty RM (1997) Natural resource endowment, the state and development strategy. J Int Dev 9 (4):651–663
Barma NH, Kaiser K, Le TM, Vinuela L (2012) Rents to riches? The political economy of natural resource-led development. The World Bank, Washington, DC
Creswell JW, Hanson WE, Clark Plano VL, Morales A (2007) Qualitative research designs: selection and implementation. Couns Psychol 35(2):236-264
Eifert B, Gelb A, Tallroth NB (2002) The political economy of fiscal policy and economic management in oil-exporting countries. Policy Research Working Paper 2899. The World Bank, Washington, DC
Gavin M (1993) Adjusting to a terms of trade shock: Nigeria, 1972-88. In: Dornbusch R (ed) Policymaking in the open economy: concepts and case studies in economic performance. Oxford University Press, New York, pp 172–219
Gelb A (1989) Oil windfalls: blessing or curse? Oxford University Press, Oxford
Gelb A (2011) Economic diversification in resource rich countries. Mimeograph, Center for Global Development, Washington, DC
Heller PRP, Marcel V (2012) Institutional design in low-capacity oil hotspots. Revenue Watch Institute, New York
Karl TL (2007) Oil-led development: social, political, and economic consequences. In: Cleveland CJ (ed) Encyclopaedia of energy. Elsevier, Amsterdam, Netherlands, pp 661–672
Kolstad I, Wiig A (2008) Political economy models of the resource curse: Implications for policy and research. CMI Working Paper WP2008:6, Chr. Michelsen Institute, Bergen, Norway
Kurdistan Regional Government (2015) [website] Available at. http://cabinet.gov.krd
Lahn G, Marcel V, Mitchell J, Myers K, Stevens P (2009) Good governance of the national petroleum sector: the Chatham House document. The Royal Institute of International Affairs, London
Lindner S (2014) Good practice in mitigating corruption risks in the extractives sector. U4 Expert Answer, Anti Corruption Resource Centre, Bergen, Norway
Luong PJ, Weinthal E (2006) Rethinking the resource curse: ownership structure, institutional capacity, and domestic constraints. Ann Rev Polit Sci 9:241–263
Marcel V (2015) Guidelines for good governance in emerging oil and gas producers. Chatham House, London
McPherson C (2003) National oil companies: evolution, issues, outlook. In: Davis JM, Ossowski R, Fedilino A (eds) fiscal policy formulation and implementation in oil producing countries. International Monetary Fund, Washington, DC, pp 184–203
McPherson C, MacSearraigh S (2007) Corruption in the petroleum sector. In: Campos JE, Pradhan S (eds) The many faces of corruption: tracking vulnerabilities at the sector level. World Bank Publications, Washington, DC, pp 191–220
Myers K, Mohammed A (2012) A short guide to parliamentary oversight of the oil and gas sector for parliament of Ghana. Natural Resource Governance Institute, New York. Available at: http://www.resourcegovernance.org/sites/default/files/PC-MPs-Short-Guide-on-Oil-Gas.pdf. Accessed 12 Feb 2016
Myers MD (2013) Qualitative research in business and management, 2nd edn. Sage Publications, Washington, DC
Robinson JA, Torvik R, Verdier T (2006) Political foundations of the resource curse. J Dev Econ 79(2):447–468
Ross M (2008) But seriously: does oil really hinder democracy? Mimeograph, UCLA Department of Political Science, Los Angeles, CA. Available at: http://leitner.yale.edu/sites/default/files/files/resources/papers/ButSeriously.pdf. Accessed 12 Feb 2016

Stevens P (2008) Kuwait petroleum corporation: searching for strategy in a fragmented oil sector. Working Paper 78, Program on Energy and Sustainable Development, Stanford University, Stanford, CA

Thurber M, Hults D, Heller PR (2010) The limits of institutional design in oil sector governance: exporting the 'Norwegian Model'. Paper presented at the international studies association annual convention, New Orleans, LA, 18 February

Thurber MC, Hults DR, Heller PR (2011) Exporting the 'Norwegian Model': The effect of administrative design on oil sector performance. Energy Policy 39(9):5366–5378

World Bank (2010) Parliamentary oversight of the extractive industries sector. World Bank Institute, Washington, DC

Yin RK (2003) Case study research: design and methods, 3rd edn. Sage Publications, Thousand Oaks, CA

Chapter 7
General Conclusions and Discussion

Abstract This chapter discusses causal explanations of the resource curse confronting most petroleum-exporting countries by investigating the political, social and economic factors through which petroleum revenues are converted into a curse rather than a blessing. The findings reveal that well-designed governance of the oil and gas sector is a key determinant of the oil and gas producing countries' success in the management of petroleum resources. The findings of the research are significant, given that few previous studies in the resource curse literature have applied a mixed methods approach. This research has shown that empirical studies are insufficient to assist petroleum exporters in addressing the transmission channels of the resource curse, whereas case studies help to highlight policy options that will deal most successfully with channels contributing to the curse. The implications of this research may be useful in helping new petroleum exporters to avoid the resource curse.

Keywords Resource curse · Petroleum revenues · Institutional policies · Well-designed institutional structure · Mixed methods approach · Kurdistan region

7.1 Introduction

The aim of this study has been to investigate the key challenges faced by petroleum-exporting countries in relation to the effective management of their petroleum wealth. The core question has been why have some hydrocarbon exporters been able to ensure that the huge revenues ensuing from oil and gas exploitation do not become a curse while others have not. To achieve the aims and objectives of this book, a mixed-methods research methodology has been employed (see Fig. 7.1) as a new way of studying the resource curse, with a view to gaining a deeper understanding of the challenges facing the governance of the oil and gas sector in petroleum-exporting countries.

K. Abdullah Auzer, *Institutional Design and Capacity to Enhance Effective Governance of Oil and Gas Wealth: The Case of Kurdistan Region*, Perspectives on Development in the Middle East and North Africa (MENA) Region, DOI 10.1007/978-981-10-4518-9_7

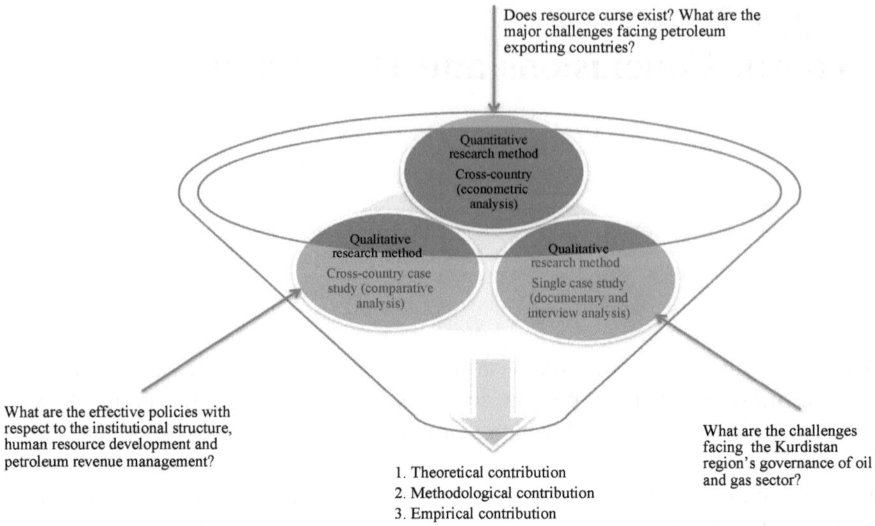

Fig. 7.1 Triangulation method

On the basis of a comprehensive literature review, the existing challenges have been classified into three categories: economic, political and social. The economic transmission challenges include Dutch disease, revenue volatility and petroleum dependency. Political and socio-economic transmission mechanisms explain the role of institutional quality and human resource capacity building in converting petroleum resources into a curse or a blessing. The study has used quantitative data to reappraise: (i) the existence of the resource curse (a negative connection between resource dependency and economic growth); (ii) whether human resource and institutional development affect growth performance positively; and (iii) the influence of natural resources on growth through institutional and human resource quality. In addition, the study has used qualitative research (cross-country case studies) to explore the effectiveness of the policies adopted by four petroleum-exporting countries (Norway, Kuwait, Azerbaijan and Nigeria) to address the economic, political and socio-economic transmission channels of the resource curse. Finally, a second source of qualitative data (semi-structured interviews) has been used to explore the perceptions of the issues surrounding the oil and gas industry in the Kurdistan region from the perspective of five target groups: the Ministry of Natural Resources, Parliament, foreign oil companies, domestic service companies and the media.

This chapter has five main objectives. First, to summarise the empirical findings on the importance of transmission mechanisms of the resource curse in resource-abundant countries and sheds light on the determinants of economic performance in natural resource-based economies, particularly in countries in which economic growth is driven almost entirely by the petroleum sector. Second, to summarise the findings relating to lessons drawn from practices of good governance of hydrocarbon resources, including institutional, human resource and revenue management policies. Third, to integrate the findings from the quantitative and qualitative data, and to highlight the implications of the findings for oil- and gas-exporting countries. Fourth, to summarise the target groups' views on the challenges facing the oil and gas sector, and to provide **policy** prescriptions for petroleum governance in the Kurdistan region. Finally, this chapter addresses the contributions and limitations of the study and offers suggestions for future research.

The chapter is organised as follows. Section 7.2 provides an overview of the book. Section 7.3 presents a summary of the literature survey findings. Section 7.4 summarises the quantitative data findings and Sect. 7.5 summarises the findings from the qualitative cross-country case studies. Section 7.6 integrates the quantitative and qualitative data and presents a general policy prescription for petroleum-exporting countries. Section 7.7 summarises the findings from the interview research. Section 7.8 provides a policy prescription for petroleum governance in the Kurdistan region. Section 7.9 describes the contributions of this study to the literature on the resource curse. Section 7.10 discusses the limitations of the study and finally, Sect. 7.11 presents suggestions for future research.

7.2 Overview of the Book

Since the 1950s, the contribution of primary products to economic growth has been a subject of controversial debate among economists (Prebisch 1950; Singer 1950; Hirschman 1958), who have argued that countries highly dependent on primary product exports face problems of declining trade and a lack of sufficient links with the rest of the economy. Since the 1970s, evidence from studies of the experiences of oil exporters has suggested that volatile oil revenues have affected the economies of oil-exporting countries (Kilian 2010).

The devastating potential effects of natural resource wealth on the economies of natural resource-abundant countries, particularly those with petroleum wealth, were first posited by Auty (1990, 1993) and Gelb (1989) as the resource curse book. They shed light on the effects of the fragile structure of economies relying heavily on oil as their major income earner. According to Auty, such economies suffer from lower growth owing to a lack of domestic economic links and mismanagement of petroleum rents. However, the resource curse book did not test his theory by applying economic modelling. Sachs and Warner (1995, 2001) first applied econometrics to examine the resource curse book in terms of explaining the apparent a negative correlation between natural resources and economic growth. They measured the economic performance of 97 developing countries for the period 1971–1989 using the ratio of natural resource exports to gross domestic product (GDP), and concluded that natural resource abundance has an adverse effect on economic growth in resource-based economies.

The emergence of a number of new oil- and gas-producing states and regions, including the Kurdistan region in the north of Iraq, has triggered renewed debate about the potential for petroleum wealth to drive economic development, as well as the possible constraints. Therefore, this study has aimed to understand why some petroleum-abundant countries have been able to translate petroleum revenues into advantageous economic development while others have not. It also seeks to devise an effective governance model for the oil and gas sector of the Kurdistan region.

Petroleum-exporting countries face many issues in terms of translating their petroleum revenues into advantageous economic development. The central aim of the studies presented in Chaps. 2 and 3 has been to identify and re-appraise the major channels through which the resource curse occurs, while the overarching aim of the studies presented in Chaps. 4–6 has been to identify effective policies adopted by petroleum-exporting countries and the challenges facing the Kurdistan region's oil and gas sector, in order to provide effective policy prescriptions for the good governance of petroleum wealth.

7.3 Summary of Challenges Facing Petroleum Rich Countries

Following Sachs and Warner's studies, a vast amount of literature has analysed the existence of the resource curse and possible channels through which natural resources may negatively impact economic growth. The various difficulties facing petroleum-exporting countries have been identified through a comprehensive survey of the relevant literature, and the existing challenges have been classified into three categories: economic, political and socio-economic.

The economic transmission challenges are based on theories of Dutch disease, revenue volatility and dependency. The most important symptoms of Dutch disease are exchange rate overvaluation, lower investment in tradable sectors, thereby obstructing industrialisation, rapid expansion of the service sector, high average wages, and unemployment (Oomes and Kalcheva 2007; Ismail 2010; Benjamin et al. 1989; Van Wijnbergen 1984).

Unstable oil markets, and thus variability of oil revenues, adversely affect economic growth in oil-exporting countries. Market instability poses a threat to investment in countries with governments that poorly formulate fiscal policy responses to volatility associated with the price of oil (Combes and Guillaumont 2002; Aizenman and Marion 1999). Furthermore, fluctuating petroleum revenues may put economic development at risk, as government expenditure tends to move pro-cyclically with total oil revenues: an oil boom increases the cost of investment, while an oil bust leads to under-spending on capital expenditure (Auty 2001; Richmond et al. 2015). For example, significant revenue shortfalls driven by the sharp decline in global oil prices since mid-2014 have impacted the economy in the major exporters of crude oil. The low oil prices have led to a budget deficit, which have financed by drawing on the saving fund. The major ill effect of this collapse in oil prices has been on the capital expenditures since the rainy day fund can protect economies in many exporting countries for a short-term. The public and private sectors have cut back on investment and thereby delaying the projects (Donnan 2016; Kerr 2016; Arezki and Blanchard 2015).

Economic over-dependence on volatile oil revenues is another determinant factor that compromises sustainable economic development in petroleum-exporting countries. The limited diversification of petroleum-rich economies results in dependency, which poses a major threat to long-term economic development in petroleum-rich countries. Diversification policies aim to move away from the oil and gas sector to other non-petroleum sectors, such as import-substitution industries that convert volatile and exhaustible petroleum incomes into other sources of income generation (Gil et al. 2014).

Beyond these economic concerns, the political mechanisms of poor economic performance refer to institutional contexts associated with the oil and gas sector. The large rents yielded by the major petroleum exporting countries give rise to corruption, which results in poor governance, which in turn enables more rent seeking activities. Government and political elites may give more opportunities and advantages to their preferred allies or networks in an environment that lacks transparency and accountability in the processes of awarding contracts, collecting revenues and royalties, distributing petroleum revenues and managing public investment. Poor institutional quality converts petroleum-abundant countries into rentier states whose major source of income is petroleum revenues. Such governments are less transparent and accountable regarding management of the petroleum sector (Karl 2004; Busse and Gröning 2013).

In addition to the economic and political channels of the resource curse, socio-economic challenges, such as a low level of human capacity building, may exacerbate the adverse effect of petroleum resources on long-term economic growth. Petroleum states under-invest in education and workforce skills because their economies are based on their endowments of petroleum resources. Such major oil-exporting countries are classic examples of a ‘rentier state’. Petroleum-led economic development may have adverse effects on human capital accumulation by neglecting the creation of conditions for improving educational quality. Human capital investment may contribute to improving educational quality, which plays a key role in diversifying income sources through skilled labour-intensive sectors, such as manufacturing and other productive activities (Shao and Yang 2014; Humphreys et al. 2007) .

7.4 Summary of the Quantitative Data Analysis of the Resource Curse

This section summarises the findings drawn from the quantitative data. Chapter 3 re-appraised the importance of the major channels through which natural resources influence economic growth, identified in the resource curse literature and discussed in detail in Chap. 2. Its main objective was not to establish causality but to infer the relative significance of the transmission channels to the resource curse. It empirically examined the existence of an adverse relationship between economies over-dependent on petroleum and economic growth, and also reappraised the positive effect of human resources and institutional capacity building on economic performance. A conditional hypobook was developed that the impact of natural resources, such as petroleum wealth, is conditional on the quality of institutions and human capital development.

Two different models were used for cross-sectional and dynamic panel data analysis. Both methods covered a sample of 160 countries over the period 1970–2010 to determine factors required for long-run economic growth. The first section used ordinary least squares (OLS) regression analysis to answer the research questions. The research focused on the direct impact of different types of natural resource that are major sources of income generation, and their indirect effect through institutional and human resources on average rates of economic growth in countries highly dependent on resource exports. Therefore, institutional and human resource quality variables and a set of further control variables were added to the regression model. These control variables were investment share of GDP to measure growth in capital stock, in other words the spending effect with respect to Dutch disease, and terms of trade to capture the effect of the volatility of petroleum revenues in countries where the petroleum sector is the major source of income generation.

As discussed in Chap. 3, the econometric analysis has shown that countries dependent on fuel, identified through the creation of dummy variables, tend to underperform in economic growth compared with less fuel-dependent peers. The evidence from the OLS estimation suggests the existence of the fuel curse in economies relying heavily on petroleum revenues. Moreover, evidence has been provided in line with the hypobook postulating that a good institutional setting will help abate the resource curse. Over the period 1970–2010, petroleum-exporting countries with higher levels of hydrocarbon dependence tended to be more exposed to the curse because of low institutional capacity. These findings corroborate the results of studies conducted on the quality of institutions as important determinants of lower growth or, in other words, the curse (Isham et al. 2005; Sala-i-Martin and Subramanian 2003; Mehlum et al. 2006; Boschini et al. 2007).

However, the results should be treated with caution since the cross-sectional analysis did not control for unobserved fixed-country effects. Furthermore, the study estimated OLS at one point in time, and therefore did not capture the effects of dynamic behaviour of the data (Levin 2006), for example the change over time of primary commodity prices such as crude oil, which have fluctuated greatly since 1970 and have a crucial effect on the economic growth of all oil-based economies.

In order to overcome the limitations of the cross-country OLS approach, the generalised method of moments (GMM) approach was used to add greater precision to the estimates. This is a dynamic framework that relates the rate of economic growth over time to changes in variables of interest. A five-year lag was used to mitigate short-run business cycle effects. In order to account for omitted invariant country characteristics such as geographical factors, country fixed effects were included; and to overcome omitted shocks occurring in all countries, such as the world business cycle effect, a period effect was added to the regression model.

The systematic GMM estimation method included a regression equation in the levels and lagged differences of the endogenous variables as an instrument to overcome the weak instruments related to variables showing little variation over time, such as institutional quality and human resource development. Furthermore, using this technique and focusing on changes in GDP over various five-year periods, it was expected that more informative results would be obtained regarding the partial effects on growth of changes in fuel, mineral and agricultural resource dependence, institutions and human resources.

The evidence of the panel data model revealed several main findings. It showed that, even controlling for fixed-effect omitted variables, higher fuel exports are related to lower long-run economic growth. The main finding associated with the effect of fuel exports on economic growth is that quality of institutions is a key factor determining the economic performance of petroleum-exporting countries. This finding is consistent with studies Karl (2005) by, Auty (1993) and Ding and Field (2005). It also shows that countries abundant in point resources are more prone to the resource curse than diffuse resource-abundant countries, which is consistent with a study by Auty (2001).

As discussed in Chap. 3, these results should be treated with caution, since limited data are available and the various dimensions of institutional quality are subject to measurement error. As Ross (2014) argues, definitions of institutional quality indices are ambiguous, and institutions may also be affected by natural resources. This makes it difficult to draw conclusions about the possible existence of a causal relationship between fuel exports and institutional quality. Moreover, how natural resources are measured may affect the results, including the distinction between natural resource dependence and natural resource abundance, and how they are gauged.

7.5 Summary of the Findings of the Qualitative Data Analysis

This section summarises comparative studies of resource-rich countries that shed light on causal explanations for poor economic growth through the transmission channels and highlight causal explanations for economic, political and social channels in which natural resource dependence may affect economic growth. These help draw policy lessons from the experience of other resource-based industries to tackle the diverse curse effects of petroleum revenues on long-run economic development. Three research questions were posed to investigate the best policy tools, addressing economic, political and socio-economic issues across the selected petroleum-producing countries:

(1) How effective are institutional design and structure measures in addressing the political and economic issues associated with accountability, transparency and checks and balances?
(2) How effective are policies and measures aimed at human resource capacity building?
(3) How effective are economic policies in dealing with macroeconomic and petroleum revenue management issues?

A comparative assessment of policy frameworks for institutional design, human resource capital development and revenue management was conducted with regard to Norway, Kuwait, Azerbaijan and Nigeria.

7.5.1 Institutional Design

A comparative assessment of the four major exporting countries has shown that institutional frameworks play a major role in maximising the outcomes of the petroleum sector. This finding is consistent with the literature suggesting that the quality of the existing institutional context determines the enhancement of value creation from petroleum wealth, which is not renewable (Barma et al. 2012). For instance, Norway has opted for an oil institutional model based on a separation of powers between the NOC, which is engaged in commercial hydrocarbon operations, the regulatory body, which provides oversight and technical expertise, and the government ministry, which helps set policy. These well-designed institutions are underlying factors that prevent conflicts of interest and have contributed to successful policy implementation in Norway's petroleum industry. Norway's oil institutional structure, with a strong emphasis on transparency and accountability, differs from those of Kuwait, Azerbaijan and Nigeria.

The findings have shown that institutional design is a fundamental factor in effective management of the petroleum sector in hydrocarbon-abundant countries. This finding is consistent with studies by Heller and Marcel (2012), Lahn et al. (2009), Thurber et al. (2010, 2011) and Luong and Weinthal (2006). Furthermore, the effective design of institutions is based on setting clear and separate goals, roles and responsibilities for the main stakeholders involved in the oil and gas sector. In this context, conflicts of interest may be reduced among the key actors: the ministry, including an independent, highly capable regulatory agency, foreign and national oil firms, and parliament. Assessment of the administrative structure of petroleum governance has indicated that best practice in petroleum sector governance is based on the separation of functions, which builds the foundation for good governance of the oil and gas sectors.

7.5.2 Developing Human Resource Capital

This sub-section summarises the findings regarding the human resource capacity-building policies adopted by the selected countries. As discussed in Chap. 3, human resource development in the oil and gas industry is lacking, and is therefore a real cause of concern and a major challenge to national oil companies (NOCs). Shortages of highly qualified and experienced workers have limited the ability of NOCs in Kuwait, Azerbaijan and Nigeria to plan and execute large-scale, complex development projects. In this context, these petroleum-exporting countries are dependent on international oil companies (IOCs) and international oilfield service companies for technical assistance. Furthermore, analysis of policies associated with developing indigenous technological capacity shows that countries relying heavily on the petroleum sector have been unable to support local skills development, capacity building and utilisation. These results confirm the findings of studies such as Glyfason (2001), Birdsall et al. (2001) that suggest that poor human capacity building contributes to the poor performance of petroleum-based economies. Some authors argue that the level of economic development at the time when the countries became an oil exporter and as well as the contribution of oil revenues to the development of human capital play a key role in the success of a few oil exporting countries, such as Norway (Lederman and Maloney 2008; Fosu 2012). Furthermore, the abundance of human capital is a major determinant in diversifying the export portfolio and thereby increasing manufacturing's share of GDP in hydrocarbon-based economies.

7.5.3 Revenue Management

This sub-section summarises the main findings with respect to effective mechanisms adopted by the selected countries to tackle the volatile and uncertain nature of petroleum revenues. It discusses effective policies that address uncertainty in petroleum revenue forecasts and possible fluctuations of oil prices. Norway, Kuwait, Azerbaijan and Nigeria have each established a petroleum revenues fund to address the negative effect of natural resources on growth through the volatility of oil prices. However, the quality of such institutional mechanisms has varied between these selected petroleum-exporting countries.

The establishment of a petroleum fund is insufficient in itself to prevent waste of revenues when the fiscal rules and procedures of the fund are ineffective. A well-designed petroleum revenue fund may impede corruption and rent seeking.

The findings have shown that the key features of the successful Norwegian petroleum fund are a consolidated budget framework, liquidity constraint on the general budget, limits on domestic investments by the fund, increased foreign investments, and transparency and accountability. These findings are consistent with studies such as those by Sturm et al. (2009), Ossowski et al. (2008) and Davis et al. (2003).

Most of the findings from the comparative cross-country study support and explain the results obtained from the statistical analysis. As discussed in Chap. 3, an oil-based economy may be vulnerable to external shocks generated by fluctuations in oil prices, but the scale of impact depends largely on the degree of concentration of petroleum exporting countries' export portfolios. Kuwait, Azerbaijan and Nigeria lack effective diversification policies to prioritise financing facilities for export-oriented industries, given that inadequate trade finance is a major constraint for petroleum exporting countries, especially small and medium-sized firms that have no access to finance.

7.6 Summary of the Findings of Interview Research

This section summarises the findings of Chap. 5 and the stakeholders' perceptions of the challenges facing the oil and gas sector presented in Chap. 6. It also provides a comparative analysis of the four selected petroleum-exporting countries and the Kurdistan region in terms of policy responses to the economic, political and socio-economic transmission channels of the resource curse.

Semi-structured interviews were used to answer three questions relating to the economic, political and socioeconomic transmission channels of the resource curse:

1. How effective, accountable and transparent are institutions in the Kurdistan region? What are the challenges and opportunities?
2. How effective and capable are domestic human resources within the Kurdistan region? What are the existing obstacles and opportunities?
3. What are the challenges currently facing economic development as a whole within the Kurdistan region?

The findings reveal institutional, social, economic and geopolitical challenges facing the effective governance of the oil and gas sector and its role in the wider economy of the Kurdistan region.

7.6.1 Institutional Challenges

The feedback obtained from the interviewees has highlighted three main challenges facing the oil and gas sector in the Kurdistan region: poor institutional design, low institutional quality, and poor safeguards and quality controls. The main finding is that its current governance suffers from a poorly-designed institutional framework. A weak policy-making body (the Regional Council) and the absence of a NOC has created a ministry-dominant model of petroleum governance. The weak regulatory function exercised by the Ministry of Natural Resources (MNR) has led to poor management of all areas in terms of timeliness, predictability and cost-effectiveness. The weakness of the regulatory system has resulted in duplication of effort, a lack of innovation and a high level of bureaucracy, which is a major obstacle facing oil companies in the Kurdistan region. The cumbersome bureaucracy is due to a lack of coordination and collaboration between the various departments in the ministry, thereby creating opportunities for corruption and rent-seeking behaviours in all petroleum activities from exploration to the refinery sector.

As participants highlighted, the lack of a well-designed governance model for petroleum wealth has resulted in increasing public concern about the accountability and transparency of the management of oil and gas wealth by the government ministry. All interviewees agreed that this is due to a lack of participation by international organisations such as EITI and the World Bank, and more importantly the ineffective role of the Kurdistan Parliament in overseeing and controlling the oil and gas sector.

The case study on petroleum governance of the Kurdistan region has shed light on the fact that institutional weakness in the petroleum sector manifests itself in institutional ineffectiveness: institutions are in place to carry out and ensure the maximisation of benefits from petroleum resources, but these are not efficiently used or enforced. Another manifestation of a weak administrative structure is the lack of necessary institutions or companies to carry out the various functions of petroleum governance effectively in order to ensure the efficient management of petroleum wealth.

Of the four countries considered in Chap. 4, in Kuwait, Azerbaijan and Nigeria institutional weakness plays a contributing role in influencing the value creation of the petroleum wealth, as in the Kurdistan region. In each case, lack of a clear administrative body leads to less transparency and accountability in governance of the oil and gas sector. Furthermore, inefficient administrative design of the governance model for the petroleum sector may result in a shortage of experienced management, insufficient technical knowledge, political and personal interference, inadequate law and regulation, duplication of efforts, excessive bureaucracy, conflicts of interest, and poor oversight and control. However, Norway's experience with its oil institutional structure is different from that of the other cases here, with a strong emphasis on transparency and accountability. Norway has opted for an oil institutional model based on a separation of powers between the NOC, which is engaged in commercial hydrocarbon operations, the regulatory body, which provides oversight and technical expertise, and the government ministry, which helps set policy.

7.6.2 Social Challenges

Most interviewees stated that governance of the oil and gas industry in the Kurdistan region has faced two key social challenges: unemployment, and ineffective management of public expectations. As explained in Chap. 6, inefficient allocation of windfalls has hampered economic development policies, such as domestic capacity building of appropriate quality and with capabilities comparable to international standards. This context makes it difficult to enhance indigenous participation in petroleum activities, use local technology, award more contracts to indigenous firms, or stimulate joint venture arrangements between indigenous and foreign companies.

In common with Kuwait, Azerbaijan and Nigeria, inadequate investment in education and human capacity building is a major challenge facing the Kurdistan region. Moreover, a noticeable mismatch between the skills supplied by the educational system and labour market demands is a challenge prevalent in these three countries as well as the Kurdistan region. Shortage of skills remains a significant barrier to productivity improvements. In contrast, Norway benefits from a competent labour force, thanks to its efficient, transparent and accountable governance structure and institutions. The terms of IOC licences make it mandatory to transfer skills and competencies to Norwegian companies, and Norwegian oil companies, in particular Statoil, set personnel training as a key objective. In addition to training schemes, the transfer of technology and the development of research by promoting cooperation between IOCs and Norwegian research institutions are successful aspects of Norway's hydrocarbon policy.

7.6.3 Economic Challenges

This sub-section examines the major problems affecting the Kurdistan region's economy. Non-transparent revenue collection and spending and continued heavy dependence on oil revenues for the prosperity of the region's people are significant obstacles to the economic development of the Kurdistan region.

The findings from Chaps. 5 and 6 have shown that Kurdistan's oil revenues have been collected and allocated by MNR to support various economic sectors. Evidence from the interview research has highlighted that petroleum revenues have not been collected and spent transparently and accountably owing to the absence of a petroleum fund. Thus, Kurdistan needs to transfer all revenues generated from the oil and gas sector into a petroleum fund in order to increase transparency and accountability in revenue management, and to save some petroleum revenues for times when oil prices are high and to smooth fiscal public expenditure when they collapse, as has been the case since mid-2014.

Unlike the Kurdistan region, Norway, Kuwait, Azerbaijan and Nigeria have adopted institutional vehicles in response to oil boom and bust. Norway has diversified its sources of income, partly by investing its petroleum revenues abroad in order to diversify risk and increase expected rates of return. This diversified investment portfolio insulates the domestic economy from Dutch disease and external shocks. The non-oil tradable sectors are gradually growing; however, the greater proportion of non-oil tradable sectors is still linked to hydrocarbon industries.

However, establishment of a petroleum fund is insufficient to address the economic challenges of the petroleum curse, such as Dutch disease and volatile oil revenues. A well-designed petroleum fund may lead to the transparent and effective management of petroleum revenues but, unlike Norway, the governments of Kuwait, Azerbaijan and Nigeria have been unwilling to impose fiscal rules on their respective funds. A lack of firm fiscal rules regarding inflows and outflows, as well as inadequate independent oversight, has resulted in arbitrary withdrawals of funds. In addition, oil funds have not been integrated into the state budgetary process, and powerful political and elite groups control transactions from the fund. Given this context, in contrast to Norway, transparency and accountability remain a major challenge for Kuwait, Azerbaijan and Nigeria.

The findings show that the Kurdistan region's economy is driven overwhelmingly by its oil and gas sector. In common with oil-dependent countries such as Kuwait, Azerbaijan and Nigeria, the region faces the adverse effects of oil boom and bust. The evidence from Chap. 5 indicates that the KRG offers few incentives to help develop other domestic economic sectors since its revenues from petroleum resources are plentiful; however, this sector is less labour intensive and has led to increases in unemployment. Furthermore, private investors are not motivated to invest in economically productive sectors, such as manufacturing and agriculture, since the service and construction sectors provide the biggest profits. Inefficient and inadequate fiscal policies, specifically a weak banking system in terms of mobilising savings, has made Kurdistan's economy more heavily dependent on petroleum revenues. Hence, the economy has become less diversified and thereby more vulnerable to external shocks.

7.6.4 Geopolitical Issues

The interviewees discussed the further challenge of the geopolitical issues facing the Kurdistan region's oil and gas sector, which are a major source of uncertainty. These include the high economic dependency of the KRG on the federal budget, the fact that it is a landlocked region, the risk of Islamic State attack, and fluctuations in oil prices.

Given that the Kurdistan region is not an independent country, the major economic challenges facing the region are difficult to address. For example, the financial dispute between Erbil and Baghdad has had the effect of delaying payments to oil and gas companies operating in the Kurdistan region, which has posed a major threat to investment plans for the oil and gas industry and other economic sectors.

However, the KRG has recently started to export its oil independently and faces the same challenges as Kuwait, Azerbaijan and Nigeria. The Kurdistan region's economy is vulnerable to oil price fluctuations. The participants' responses shed light on the fact that the Kurdistan region's dependence on oil revenues is a major challenge to long-term economic development. This has occurred as a result of poor financial policy and ineffective petroleum revenue management.

Another key geopolitical issue outlined by interviewees is linked to petroleum exports. They explained that the Kurdistan region is landlocked, which raises issues of infrastructural dependency in exporting its oil and gas. The expansion of pipeline exports is the lifeblood of the Kurdistan economy, so the future of landlocked Kurdistan depends on political relations between Ankara and Erbil.

Another geopolitical issue highlighted by participants is security, which plays a vital role in the ongoing development of the petroleum sector. Security is important for optimal and long-term investments and employment development, which have been hit particularly badly since companies have been limiting their commitments as a result of insurgent attacks.

7.7 Policy Prescription for Petroleum Governance in the Kurdistan Region

The importance of the oil and gas sector to the sustainable development of the Kurdistan region's economy has been highlighted in previous chapters. The findings of this book have significant institutional consequences for the Kurdistan region in that they focus on measures that to enable the KRG to protect the institutional and human resource capacity-building base of the petroleum sector, and the region at large. The findings also indicate that, as a result of an ineffective governance model for the oil and gas industry, the Kurdistan region has failed to manage its petroleum revenues effectively. Therefore, as a result of this research, the following governance structure for the oil and gas sector in the Kurdistan region is proposed:

As shown in Fig. 7.2, the suggested governance model has three main pillars—policy formulator, regulator and operator—to prevent conflicts of interest and manage petroleum revenues in a more transparent and accountable way. The policy pillar should hold overall responsibility for the management of petroleum resources, focusing on oil and gas policy issues such as opening new areas for exploration, allocating contracts and approving development plans for awarded blocks. The Supreme Petroleum Council (SPC) should be responsible for policy setting. A clear understanding of the responsibilities and organisational structure of the SPC may lead to greater accountability and transparency in the vision, policies and plans of the KRG for its petroleum resources.

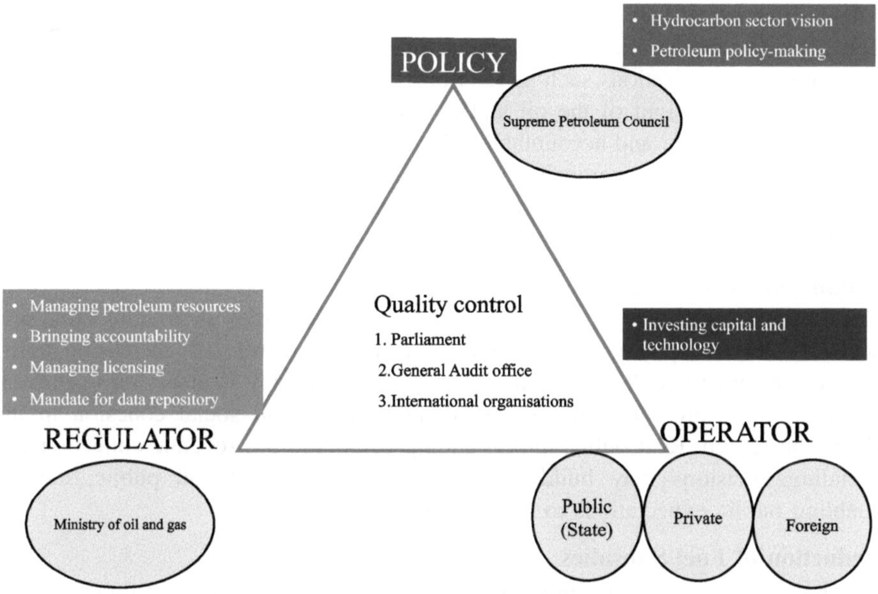

Fig. 7.2 Proposed governance model for Kurdistan region's oil and gas sector

Monitoring by the Oil and Gas Ministry

The Oil and Gas Ministry should regulate, monitor and supervise all petroleum activities implemented by independent entities, including the Kurdistan Exploration and Production Company (KEPCO), the Kurdistan National Oil Company (KNOC), the Kurdistan Oil Marketing Organisation (KOMO) and the Kurdistan Organisation for Downstream Operations (KODO), as well as foreign oil companies. The government ministry should be the driving force in realising the region's petroleum resource potential by emphasising long-term solutions, upside opportunities, local content development, cooperation with other authorities and comprehensive follow-ups of petroleum activities. It should have national responsibility for ensuring that the state collects revenues, taxes and fees from the petroleum sector, collects and analyses petroleum data, conducts audits and publishes consistent data on petroleum activities. The Ministry of Oil and Gas should also act as the SPC's technical arm and advisory body.

Involvement of NOCs

National Oil Companies need to be involved in petroleum sector activities in order to localise the oil and gas industry by transferring technical capability within the sector and thus building indigenous capacity. Foreign oil companies, as investors, need to deliver corporate social responsibility programmes that provide sufficient benefit to local populations through hiring and training nationals, creating a national supply and investing in infrastructure (such as capturing flared gas and developing a gas network) and ancillary industries such as refineries.

Effective Oversight by Independent Bodies

The effective role of parliament, the independent general auditor's office and related international organisations such as IETI and the Revenue Watch Institute is crucial in exercising oversight of the oil and gas sectors and thereby making petroleum activities transparent and accountable. Members of the natural resource committee need to be educated specifically on oil issues to fill the current communication gap. In addition, the general auditor's office needs to improve its technical capacity in terms of auditing petroleum activities.

Establishment of a Petroleum Fund

Regarding petroleum revenue management, a well-designed Petroleum Fund and access to detailed payment information play important roles in increasing transparency in revenue collection and spending. Transparency is also vital to prevent rumours from spreading, thus encouraging political and social cohesion in the Kurdistan region. Real public understanding of oil and gas revenue collection and spending decisions may build trust between government and public, thereby enabling public expectations to be managed more effectively.

Reduction of Fuel Subsidies

The KRG has made little effort to raise non-oil revenues. A huge proportion of government expenditure has been allocated to maintaining fuel subsidies, which benefit richer households disproportionately because the quantity of petroleum products consumed by poorer households is lower. This inequality could be mitigated by removing the subsidies, thereby directly supporting low-income households and financing better targeted projects to compensate the poorest households.

Total subsidies to the electricity sector by the KRG cost $3 billion a year, so major reforms are needed to reduce subsidies to the power sector. Replacing diesel-fired power stations with natural gas-fired electric power plants, reforming the level and structure of electricity prices, and increasing private-sector participation in the electricity sector are effective strategies to eliminate subsidies while enhancing the government's capacity for capital expenditure. Therefore, in order to raise non-oil revenues over the medium term, the KRG should redirect expenditure away from such subsidies. This may help finance greater public investment in non-oil sectors of the economy and also finance poor households in order to reduce poverty and inequality.

Prudent Investment in Other Sectors

In order to convert depleting petroleum wealth into other assets, it is vitally important to invest oil and gas revenues prudently. Given the Kurdistan region's insufficient technical education and infrastructure, it may be difficult for it to move directly to industrial production. Development of the agricultural sector is vital to the region's medium-term future. In particular, investment of oil revenues in enhancing the productivity of the agricultural sector may reduce poverty and enable broader industrial economic development. Import substitution policies would build up the Kurdistan region's non-petroleum exporting sector and eventually enable it to become self-sufficient in major food production. Furthermore, the private sector should be encouraged to engage in food production and agro-processing.

Development of Small and Medium-Sized Enterprises (SMEs)
In order to address the crowding-out effects relating to Dutch disease, it is vital to build an effective fiscal fund to help promote small and medium-sized enterprises in the Kurdistan region and thereby support economic growth and development. Since SMEs are labour intensive, they may play a major role in increasing employment, developing skills and alleviating poverty.

Strengthening the Financial Sector
The weak domestic banking system is unable to mobilise savings and channel them effectively into private investment. Therefore, in the long term, building the capacity of the financial sector will be critical in mobilising resources for domestically generated growth in the Kurdistan region.

Reform of the Tax System
The existing tax system in the Kurdistan region is limited and is ineffective in meeting rising budget financing needs in the context of a period of low oil prices.

Investment in Human Capital
Human capital investment must be increased in order to educate and train local experts and skilled labour adequately. Development of the education system (higher education and vocational institutions) may contribute positively to increasing participation in the industry, in particular in oil and gas.

Diversification of Petroleum Trade Routes
The Kurdistan Region is landlocked and far from international oil markets. Lack of territorial access to the coast makes the region dependent mainly on pipelines to transport its hydrocarbon to world markets and limits its petroleum trading activities with the rest of the world. Major crude oil export pipelines include the pipeline from the Kurdistan Region to Turkey's Ceyhan port, which poses a threat of high dependency on Turkish trade routes. The second export route for crude oil is through Iran, trucking Kurdish oil to Iran's Bandar Imam Khomeini for shipping to Asia.

For the Kurdistan Region, being landlocked poses distinct challenges, particularly with respect to trade. In addition to transportation costs, the quality and reliability of transportation is dependent on factors outside the sphere of influence of the Kurdistan Region. Given these challenges, it is crucial that the KRG seeks to develop its trade and transport routes and reduce its high dependency on petroleum exports through diversification of petroleum trade routes, such as the development of local power generation and electricity exports.

7.8 Integration Between Quantitative and Qualitative Data and a General Policy Prescription for Petroleum-Exporting Countries

The previous sections have presented separate summaries of the findings from the quantitative and qualitative data. As discussed in Chap. 1, this study has employed a mixed-methods approach, which is very useful for explaining the results when both quantitative and qualitative data are used (Creswell and Clark 2011; Boyd et al. 2012). The mixed method is uncommon in the resource curse literature. Having first obtained statistical findings with respect to the transmission of the resource curse, this study has used an explanatory sequential design (Creswell and Clark 2011) to gain deeper insights from cross-country comparative analysis.

The main findings of the cross-country comparative analysis support the results obtained from the statistical analysis. Best practice in petroleum wealth management indicates that the petroleum resource curse is not inevitable. Prudent and sound institutional, human resource capacity building and petroleum revenue management policies help avoid transmission of the resource curse. Although most of the literature on the resource curse has paid little attention to the role of administrative design in the oil and gas sector as a key factor in increasing the value creation of petroleum wealth, this study suggests that a well-designed institutional governance model for the petroleum industry is vital to addressing common problems associated with effective economic and human resource development.

The main findings suggest the following recommendations for policy makers in petroleum-exporting and newly-emerging petroleum-rich countries for the effective governance of their oil and gas sectors.

Governance of Petroleum Wealth

- Separation of policy-making, regulatory and commercial functions to avoid conflicts of interest and increase transparency and accountability.
- Establishment of NOC as a tool to achieve wider socio-economic policy objectives, such as economic diversification and raised local educational levels.
- An independent NOC board to reduce political interference in the decision-making process and delays in the approval process, as well as enabling the NOC to maximise its limited technical capacity and become more transparent and accountable.
- Enhancement of the technical capability of the regulatory body.
- Improvement of parliamentary oversight and control.

Human Resource Capacity

- Efficient local content measures to increase levels of productivity and competitiveness in oil- and gas-exporting countries.
- Localisation of the workforce and the development of local servicing companies.

- Enhancement of the quality of educational and vocational systems.
- Staff training and a constant upgrading of workers' skills.
- Effective use of petroleum revenues.

Petroleum Revenue Management

- A well-functioning petroleum fund through:
  - A consolidated budget framework.
  - Liquidity constraints on the general budget.
  - Limits on domestic investment by the fund.
  - Investing petroleum revenues abroad.
- Diversification of export portfolios through:
  - Creating links within the petroleum sector, such as petrochemical manufacturing and service sectors.
  - Creating links with other economic sectors, such as agriculture and industrial manufacturing.

7.9 Contributions of the Research

This research has critically engaged with the resource curse literature by adopting a triangulation approach through cross-country quantitative analysis (econometric analysis), cross-country case analysis (comparative analysis) and a single case study (documentary research and interview analysis). The research represents a critical reappraisal of the resource curse book. It extends current knowledge and takes an important step towards explaining why some petroleum-exporting countries successfully develop their economies whilst others fail, as well as identifying the main challenges facing the governance of Iraqi Kurdistan's oil and gas wealth. This research makes theoretical, methodological and empirical contributions to the literature on the resource curse in general, and on the Kurdistan region's oil and gas sector in particular.

7.9.1 Theoretical Contributions

From a theoretical point of view, this study has reappraised the resource curse theory, and has identified the main transmission channels of the resource curse in petroleum-exporting countries. It contributes theoretically by extending the application of political economy theory to examine empirically the role of institutional and human resource capacity building in sustainable economic development. It has focused particularly on the managerial model of the petroleum sector, in the context of which few previous studies have investigated the effect of administrative design as a causal factor affecting performance. An important theoretical contribution of this research, which has not been investigated by previous studies, is its empirical investigation of the governance model of the Kurdistan region, as a newly-emerging petroleum-exporting region. This study has assessed the effectiveness of its management of the oil and gas sector, and the role of the petroleum industry's governance model in the capacity building and development of other economic sectors.

7.9.2 Methodological Contributions

This research has provided valuable methodological insights that may be incorporated into future research in the same field. It has overcome the methodological limitations of previous studies, which have been confined mainly to the cross-country quantitative method and detailed qualitative case studies to examine the resource curse theory. This research has addressed the issues from several directions by adopting a methodological triangulation approach. In addition to conducting econometric analysis to identify the main causal mechanisms through which petroleum rents have an impact on economic outcomes, a case study analysis has been carried out to provide a causal explanation of the potential implications of hydrocarbon rents for sustainable economic development in petroleum-exporting countries.

A further methodological contribution relates to understanding the development of the newly-emerging oil and gas industry in the Kurdistan region by conducting interview research based on the results of the quantitative and qualitative cross-country studies. This method has provided an in-depth explanation of the impact of the governance model on sustainable economic development in the Kurdistan region as a new player in global energy security.

7.9.3 Empirical Contributions

There have been few previous studies in the field of petroleum sector governance models. The methods and findings of this research offer a basis for political economy scholars to expand studies of the political transmission channels of the resource curse. The research has revealed that the petroleum sector's governance model is a significant factor in promoting or hindering the effective management of oil and gas resources by the KRG. The findings of this research may assist policy makers in redesigning the administrative structure of the oil and gas sector in the Kurdistan region. In addition, the approach adopted by this study might be used to study other regions and states with similar circumstances or characteristics.

7.10 Limitations of the Research

This research has aimed to investigate the main challenges facing petroleum-exporting countries from a broader socio-economic perspective. As with all studies, however, confidence in the findings must be considered in light of the limitations. Limited data are available, and variables such as various dimensions of institutional quality and economic growth are subject to measurement error.

7.11 Recommendations for Future Research

This research has provided a better understanding of the issues facing petroleum-exporting countries by addressing them from different directions. Since few studies have applied the triangulation approach, it is recommended that further such studies be conducted to investigate the best policies to achieve better performance over the long run, and to identify the main transmission channels of the resource curse.

In addition, the results of this study show that an effective governance structure for the oil and gas sector will create the conditions for sustained economic growth and improve human resource capacity building and institutional quality, leading to a lower level of corruption and rent-seeking activities. It is recommended that future studies should examine the effect of administrative design of the oil and gas sector on the wider economy of petroleum-exporting countries, in particular newly emerging petroleum exporters.

There are a number of additional areas for further research that have been highlighted by this book. These include the further investigation of the importance of different transmission channels of the resource curse by using different regression models such as switching regression models. Switching regression models allow one to take into account country heterogeneity by looking at effects common for a group of countries. The approach could also provide a threshold value to classify natural resource dependency as a blessing or a curse.

An alternative way of using institutional variables is to create a composite index for the institutional indicators, which are highly correlated with each other, but not correlated with other covariates. The creation of a composite index would reduce the problems of collinearity and confounded effects. The disadvantage is that the effects of the different indicators are distinguished in the index estimation using principal component analysis and factor analysis, but are not separated in the regression model.

Moreover, the squares of the explanatory variables could be added to the regression to measure additional nonlinear effects. Using squared explanatory variables allows the computation of the optimal level of natural resources as a share of GDP. The optimal level will differ across countries and over time.

Another alternative approach to estimate the effects of natural resources on economic growth is to use matching techniques. In this case, the resource rich countries are compared with counterfactuals, namely countries with identical characteristics but without natural resources. The explanatory variables are measured with errors so they are often approximation. Measurement errors in dependent variables are absorbed in the residual term. However, measurement errors in the explanatory variables can lead to biased parameter estimates. Secondary data suffer from both aggregation and approximation bias sources, and thus measurement errors normally create attenuation bias. This implies that the empirical results in Chap. 3 may underestimate the impact of natural resource on economic growth. A possible solution to this problem is the use of an instrumental variable estimator, although it is normally hard to find an adequate instrument in typical economic datasets. Future research could use the estimation result in forecasts and policy simulation predicting policy outcomes for the emerging oil and gas sectors, and could also apply the methodology employed in this book to other interesting case studies, such as Scotland or Alberta.

References

Aizenman J, Marion N (1999) Volatility and investment: interpreting evidence from developing countries. Economica 66(262):157–179

Arezki R, Blanchard O (2015) The 2014 oil price slump: seven key questions. VOX, 13 Jan. Available at: http://www.voxeu.org/article/2014-oil-price-slump-seven-key-questions. Accessed 12 Feb

Auty RM (1990) Resource-based industrialization: sowing the oil in eight developing countries. Oxford University Press, Oxford

Auty RM (1993) Sustaining development in mineral economies: the resource curse book. Routledge, London

Auty RM (ed) (2001) Resource abundance and economic development. Oxford University Press, Oxford

Balouga J (2012) Nigerian local content: challenges and prospects. Int Assoc Energy Econ 6 (3):23–26

Benjamin NC, Devarajan S, Weiner RJ (1989) The 'Dutch' disease in a developing country: oil reserves in Cameroon. J Dev Econ 30(1):71–92

Birdsall N, Pinckney T, Sabot R (2001) Natural resources, human capital and growth. In: Auty RM (ed) Resource abundance and economic development. Oxford University Press, Oxford, pp 57–75

Boschini AD, Petterson J, Roine J (2007) Resource curse or not: a question of appropriability. Scand J Econ 109(3):593–617

Boyd B, Santos M, Shen W (2012) International developments in executive compensation. Corp Gov: Int Rev 20(6):511–518

Busse M, Gröning S (2013) The resource curse revisited: governance and natural resources. Public Choice 154(1–2):1–20

Barma NH, Kaiser K, Le TM, Vinuela L (2012) Rents to riches? The political economy of natural resource-led development. The World Bank, Washington, DC

Combes JL, Guillaumont P (2002) Commodity price volatility, vulnerability and development. Dev Policy Rev 20(1):25–39

Creswell JW, Clark VLP (2011) Designing and conducting mixed methods research, 2nd edn. SAGE Publications, London

Davis J, Ossowski R, Daniel JA, Barnett S (2003) Stabilisation and savings funds for non-renewable resources: experience and fiscal policy implications. In: Davis JM, Ossowski R, Fidelino A (eds) Fiscal policy formulation and implementation in oil-producing countries. International Monetary Fund, Washington, DC, pp 273–315

Ding N, Field BC (2005) Natural resource abundance and economic growths. Land Econ 81 (4):496–502

Donnan S (2016) Falling oil price raises prospect of bailouts for crude producers. The Financial Times, Jan 28. Available at: http://www.com/intl/cms/s/0/c942a4c2-c5df-11e5-b3b1–7b2481276e45.html#axzz404BRd3e8. Accessed 14 Feb 2016

Fosu AK (ed) (2012) Development success: historical accounts from more advanced countries. OUP Oxford

Gelb A (1989) Oil windfalls: blessing or curse? Oxford University Press, Oxford

Gil SI, Izvorski I, van Eeghen W, De Rosa D (2014) Diversified development: making the most of natural resources in Eurasia. The World Bank, Washington, DC

Gylfason T (2001) Natural resources, education, and economic development. Eur Econ Rev 45(4–6):847–859

Heller PRP, Marcel V (2012) Institutional design in low-capacity oil hotspots. Revenue Watch Institute, New York

Humphreys M, Sachs JD, Stiglitz JE (2007) Introduction: what is the problem with natural resource wealth? In: Humphreys M, Sachs JD, Stiglitz JE (eds) Escaping the resource curse. Columbia University Press, New York, pp 1–20

Hirschman A (1958) The strategy of economic development. Yale University Press, New Haven

Isham J, Woolcock M, Pritchett L, Busby G (2005) The varieties of resource experience: natural resource export structures and the political economy of economic growth. World Bank Econ Rev 19(2):141–174

Ismail K (2010) The structural manifestation of the Dutch disease: the case of oil exporting countries. IMF Working Paper No. 10/103, International Monetary Fund, Washington, DC

Karl TL (2004) Oil-led development: social, political, and economic consequences. In: Cleveland CJ (ed) Encyclopaedia of energy. Elsevier, Amsterdam, Netherlands, pp 661–672

Karl TL (2005) Understanding the resource curse. In: Tsalik S, Schiffrin A (eds) Covering oil: a reporter's guide to energy and development. Open Society Institute, New York, pp 21–26

Kilian L (2010) Oil price volatility: origins and effects. WTO Staff Working Paper ERSD-2010-02, World Trade Organization, Geneva, Switzerland

Kerr S (2016) Gulf: IMF urges Gulf countries to reduce spending as oil price plunges. The Financial Times, Jan 21. Available at: http://www.ft.com/intl/cms/s/0/a763aeba-a0b9-11e4-9aee-00144feab7de.html#axzz404BRd3e8. Accessed 14 Feb 2016

Lahn G, Marcel V, Mitchell J, Myers K, Stevens P (2009) Good governance of the national petroleum sector: the Chatham House document. The Royal Institute of International Affairs, London

Lederman D, Maloney WF (2008) In search of the missing resource curse. Mimeograph, The World Bank, Washington, DC

Levin KA (2006) Study design III: cross-sectional studies. Evid Based Dent 7(1):24–25

Luong PJ, Weinthal E (2006) Rethinking the resource curse: ownership structure, institutional capacity, and domestic constraints. Ann Rev Polit Sci 9:241–263

Marcel V (2015) Guidelines for good governance in emerging oil and gas producers. Chatham House, London

Mehlum H, Moene K, Torvik R (2006) Institutions and the resource curse. Econ J 116(508):1–20

Ossowski R, Villafuerte M, Medas PA, Thomas T (2008) Managing the oil revenues boom: the role of fiscal institutions. International Monetary Fund, Washington, DC

Oomes N, Kalcheva K (2007) Diagnosing Dutch disease: does russia have the symptoms? IMF Working Paper 07/102, International Monetary Fund, Washington, DC

Prebisch R (1950) The economic development of Latin America and its principal problems. (Reprinted in Econ Bull Lat Am 7(1):1–22, 1962)
Richmond C, Yackovlev I, Yang SCS (2015) Investing volatile resource revenues in capital-scarce economies. Pacific Econ Rev 20(1):193-221
Ross ML (2014) What have we learned about the resource curse? Mimeograph, UCLA Department of Political Science, Los Angeles, CA
Sachs DJ, Warner AM (1995) Natural resource abundance and resource growth. Working Paper 5398, National Bureau of Economic Research, Cambridge, MA. Available at: http://www.nber.org/papers/w5398. Accessed 12 Feb 2016
Sachs JD, Warner AM (2001) The curse of natural resources. Eur Econ Rev 45(4):827–838
Sala-i-Martin X, Subramanian A (2003) Addressing the natural resource curse: an illustration from Nigeria. IMF Working Paper 03/139, International Monetary Fund, Washington, DC
Shao S, Yang L (2014) Natural resource dependence, human capital accumulation, and economic growth: a combined explanation for the resource curse and the resource blessing. Energy Policy 74:632–642
Singer HW (1950) US foreign investment in underdeveloped areas: the distribution of gains between investing and borrowing countries. In: American economic review, papers and proceedings, vol 40, pp 473–485
Sturm M, Gurtner F, Alegre GJ (2009) Fiscal policy challenges in oil-exporting countries: a review of key issues. ECB Occasional Paper 104, European Central Bank, Frankfurt, Germany
Thurber M, Hults D, Heller PR (2010) The limits of institutional design in oil sector governance: exporting the 'Norwegian Model'. Paper presented at the international studies association annual convention, New Orleans, LA, 18 Feb
Thurber MC, Hults DR, Heller PR (2011) Exporting the 'Norwegian Model': the effect of administrative design on oil sector performance. Energy Policy 39(9):5366–5378
Van Wijnbergen S (1984) 'The Dutch disease': a disease after all? Econ J 94(373):41–55

Author Index

K. Abdullah Auzer, *Institutional Design and Capacity to Enhance Effective Governance of Oil and Gas Wealth: The Case of Kurdistan Region*, Perspectives on Development in the Middle East and North Africa (MENA) Region, DOI 10.1007/978-981-10-4518-9

Subject Index

K. Abdullah Auzer, *Institutional Design and Capacity to Enhance Effective Governance of Oil and Gas Wealth: The Case of Kurdistan Region*, Perspectives on Development in the Middle East and North Africa (MENA) Region, DOI 10.1007/978-981-10-4518-9

MIX
Papier aus verantwortungsvollen Quellen
Paper from responsible sources
FSC® C105338
FSC
www.fsc.org

If you have any concerns about our products,
you can contact us on
ProductSafety@springernature.com

In case Publisher is established outside the EU,
the EU authorized representative is:
Springer Nature Customer Service Center GmbH
Europaplatz 3, 69115 Heidelberg, Germany

Printed by Libri Plureos GmbH
in Hamburg, Germany